Electrical Applications 2

Electrical

Applications 2

David W. Tyler, CEng, MIEE

Formerly Senior Lecturer, Electrical Engineering
Reading College of Technology

BH NEWNES

Newnes
An imprint of Butterworth-Heinemann Ltd
Linacre House, Jordan Hill, Oxford OX2 8DP

℟ A member of the Reed Elsevier group

OXFORD LONDON BOSTON
MUNICH NEW DELHI SINGAPORE SYDNEY
TOKYO TORONTO WELLINGTON

First published 1987
Reprinted 1990 (twice), 1991
Revised and reprinted 1993
Reprinted 1994

British Library Cataloguing in Publication Data
Tyler, D. W.
  Electrical applications
  2,
  1. Electric engineering
  I. Title
  621.3   TK145

ISBN 0 7506 0525 1

Printed in Great Britain by The Bath Press, Avon

Contents

Preface

This book covers the BTEC NII level objectives in Electrical Applications U86/330. To understand the applications, a knowledge of the underlying principles is needed and these are covered briefly in the text.

Included in each chapter are worked examples which should be carefully worked through before progressing to the next section. At the ends of chapters, further problems are provided for consolidation and self testing; where these have numerical answers, they may be found at the end of the book. In a subject such as this, many problems ask for explanations and descriptions and here the answers must be sought in the text. When dealing with a descriptive question, a good diagram almost always helps to give a clear answer and saves many words of explanation. The book aims to promote this approach by the use of over 170 figures throughout the eight chapters.

I wish to thank the Institution of Electrical Engineers for permission to quote from the Regulations for Electrical Installations. Any interpretation placed on these regulations is mine alone. In reprinting the book the opportunity has been taken to bring up to date the sections dealing with the IEE Regulations and other regulations that concern safety, electricity supply and electrical installations.

In thise days of ever-increasing use of computers, there is a growing tendency to regard any piece of equipment without a visual display as being in some way deficient. I would like to remind those who have this impression that, without power equipment and 'power men', the light current side of the industry could not exist.

I wish all readers of this book well with their studies.

David W. Tyler
October 1992

1 Transmission and distribution of electrical energy

Aims: At the end of this chapter you should be able to:

Explain why transmission is carried out at very high voltages.
Understand the factors which affect the design and arrangement of the transmission and distribution system.
Compare overhead lines with underground cables.
Explain the purpose of switchgear.
Describe the equipment and layout of a small distribution sub-station.
Describe typical three-phase industrial installations.
Calculate the current distribution in, and the efficiency of, radial feeders and ring mains.

SYNCHRONOUS GENERATORS

Virtually all the generation of electrical energy throughout the world is done using three-phase synchronous generators. Almost invariably the synchronous generator has its magnetic field produced electrically by passing direct current through a winding on an iron core which rotates between the three windings or phases of the machine. These windings are embedded in slots in an iron stator and one end of each winding is connected to a common point and earthed. The output from the generator is taken from the other three ends of the windings. The output from a three-phase generator is therefore carried on three wires. In many three-phase diagrams single line representation is used when each line on the diagram represents three identical conductors. *Figure 1.1* is drawn using this method.

All such generators connected to a single system must rotate at exactly the same speed, hence the term synchronous generator.

They are driven by prime movers using steam generated by burning coal or oil or by nuclear reactors, water falling from a higher to a lower level, or aircraft gas turbines burning oil or gas. A very small amount of generation is carried out using diesel engines.

Generators range in size from 70 MVA (60 MW at 0.85 power factor) at a line voltage of 11 kV which were mostly intalled in the 1950s, through the intermediate size of 235 MVA (200 MW at 0.85 power factor), to the recent machines rated at 600 MVA (500 MW) which generate at 25.6 kV. There are generators rated at 660 and 1000 MW but these are rare at the moment.

ECONOMICS OF GENERATION AND TRANSMISSION

The power in a single phase circuit = $VI \cos \phi$ watts where V and I are the r.m.s. values of circuit voltage and current respectively and ϕ is the phase angle between the current and voltage.

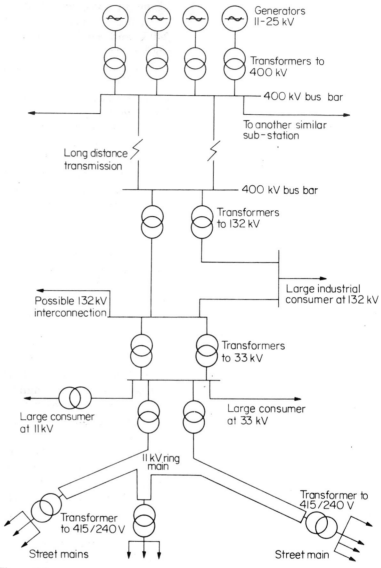

Figure 1.1

As an example, consider a power of 1 MW at 240 V and a power factor of 0.8 lagging. (1 MW = 1000 kW = 10^6 watts.)

$$240 \times I \times 0.8 = 10^6$$

$$I = \frac{10^6}{240 \times 0.8} = 5\,208 \text{ A}$$

By increasing the voltage to, say, 20 000 V the required current falls to 62.5 A.

The voltage drop in a transmission line due to the resistance of the line = IR volts.

The power loss = voltage drop × current flowing
$$= IR \times I$$
$$= I^2R \text{ watts.}$$

Using the above values of current it may be deduced that:

1 for a conductor of given size and resistance, the line losses at 240 V and 5208 A will be very much greater than at the higher voltage; or

2 if the losses are to be the same in both cases the conductor for use at 240 V will need to have a very much lower resistance and hence have a much larger cross-sectional area than that for use at the higher voltage.

To enable large powers to be transmitted through small conductors while keeping the losses small therefore requires the use of very high transmission voltages. The voltages commonly in use in the UK are shown in *Figure 1.1*. At each stage circuit breakers would be employed but these are not shown for simplicity.

Each generator feeds directly into a step-up transformer which increases the voltage to 400 kV. Power is transmitted to the major load centres at this voltage where it is transformed down to 132 kV (sometimes an intermediate step at 275 kV is used). Some heavy industry is fed at this voltage but most of the 132 kV system forms local distribution to 33 kV substations. These feed industry and a series of 11 kV substations. Ring mains at 11 kV feed transformers which supply power at 415/240 V to domestic and commercial consumers.

To save money on transformers it would seem that generation could best be carried out at 400 kV but so far it has not been found possible to develop insulation for use in rotating machines which will withstand such a high voltage while allowing the heat produced in the winding to be dissipated. In particular there are problems where the conductors leave the slots in the iron core and emerge into the gas filled spaces at the ends of the stator. The problems are overcome in transformers for use at 400 kV by the use of paper insulation and immersing the windings and core completely in a special oil which insulates electrically and convects heat away.

Cables for extra-high-voltage work are also paper insulated and contain oil under pressure. They are laid individually and heat is conducted away by the soil.

As the voltage is increased, as we have already seen, the size and hence cost of the conductors decreases. However as the voltage is increased the cost of insulation is increased. Cable insulation becomes thicker, oil is used and this must often be maintained under pressure which requires additional plant. Very expensive cable terminations called *sealing ends* have to be used.

Switchgear for use at high voltages is more complicated, bulkier and more expensive than that for use at medium and low voltages. When a circuit breaker opens to interrupt a circuit an arc is drawn between the contacts. At domestic voltages the arc is small and arc extinction occurs quickly in the atmosphere. At extra-high voltages the arc is much more difficult to extinguish and air or oil often under pressure have to be used. In addition, all the electrical parts must be kept well away from earth and these clearances are much greater where very high voltages are used.

The capital costs of extra-high-voltage gear reflect voltage levels but are not affected very much by the cross-sectional area of the conductors used.

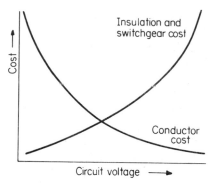

Figure 1.2

Figure 1.2 shows comparative costs of conductors and insulation for increasing system voltages.

In addition to the cost of equipment there is the provision of land to consider. The additional bulk of extra-high-voltage gear means that substations may occupy land areas of hundreds or even thousands of square metres.

In *Figure 1.1* we see that major transmission is at 400 kV. Since the power transmitted from a single power station may be in excess of 2 000 MW, the high cost of insulation and switchgear is justified by the considerable reduction in conductor costs. At distribution level the supply for a single factory or for housing from an individual transformer represents a relatively small power so that even at much lower voltages the current involved is quite small. The cost of extra-high-voltage switchgear would not be justified and the land area for a substation might well be restricted.

As local demands decrease the voltage at which they are supplied is reduced. A large factory requiring 100 MW will be fed directly from either the 132 kV or 33 kV system. A smaller factory requiring only 1 MW could be fed from the 11 kV system whilst a group of houses and shops with a collective requirement of 500 kW will be fed at 415/240 V. The conductor cross-sectional areas generally lie between 225 mm^2 and 650 mm^2 irrespective of voltage, the insulation and switchgear costs and the land area per substation decreasing at each successive voltage reduction.

Typical transformer ratings at the various voltage levels are:

25.6/400 kV	600 MVA
400/132 kV	150–250 MVA
132/33 kV	50–75 MVA
33/11 kV	10–15 MVA
11 kV/415/240 V	250–500 kVA

OVERHEAD LINES

Overhead lines for power transmission are almost invariably made of aluminium with a steel core for strength. The bare conductors are supported on insulators made of porcelain or glass which are fixed to wooden poles or steel lattice towers.

Figure 1.3 shows some typical British line supports together with the associated insulators. All the steel lattice towers shown use suspension insulators whilst the wooden poles may use either type. Three conductors comprise a single circuit of a three-phase system so that the 33 kV single circuit tower has three cross arms and three suspension insulators. Towers with six cross arms carry two separate circuits.

On high-voltage lines each support must carry a consecutive recognition number and a circuit identifying colour disc. The supports must be capable of supporting the line without movement in the ground when both line and supports are carrying a specified ice loading and an 80 km/hour wind is blowing. Safety factors of 2.5 for steel towers and 3.5 for wood must be allowed. A safety factor of 2.5 means that with the ice and wind loading the load is 1/2.5 of that which would cause the support to collapse.

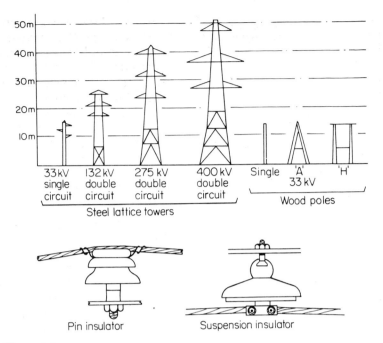

Figure 1.3

Wood supports are red fir impregnated with creosote and may be in the form of single poles or two poles made into an A or H. In the UK they are used for circuits up to 33 kV but in other countries lines up to 250 kV using 50 m poles have been erected. Since Britain imports most of the trees required and each pole is in fact a complete tree trunk, large ground clearances using poles proves to be extremely expensive.

Towers are made of steel angle section and may easily be fabricated up to almost any height by adding extra bottom sections or trestles.

COMPARISON BETWEEN OVERHEAD LINES AND UNDERGROUND CABLES

Cost The overhead line is air insulated and is supported on insulators mounted on towers or poles which are 100–400 m apart. The underground cable is fully insulated and armoured to protect it against mechanical damage and then covered overall with a corrosion resistant material.

For extra-high voltage work the overhead line is made of steel cored aluminium while the underground cable is made of copper to reduce the resistance of a given cross-sectional-area cable. Local underground distribution cables may use solid aluminium cores which are insulated with p.v.c. Four such cores are laid up, armoured and served overall to form a three-phase cable with the

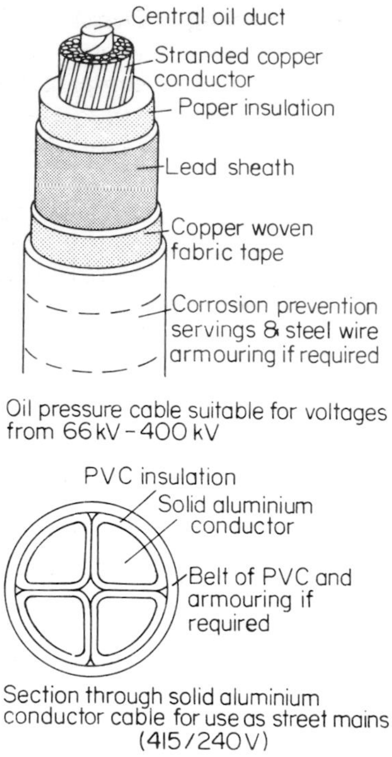

Central oil duct
Stranded copper conductor
Paper insulation
Lead sheath
Copper woven fabric tape
Corrosion prevention servings & steel wire armouring if required

Oil pressure cable suitable for voltages from 66 kV – 400 kV

PVC insulation
Solid aluminium conductor
Belt of PVC and armouring if required

Section through solid aluminium conductor cable for use as street mains (415/240 V)

Figure 1.4

fourth conductor as neutral or earth connection. This is shown in *Figure 1.4* together with an oil pressure cable, three of which are required to form a three-phase circuit.

The high cost of copper, insulation, armouring and corrosion protection, together with that of taking out a suitable trench and refilling it, make the e.h.v. underground cable many times more expensive than the overhead line. The price difference at 415/240 V using aluminium cables is not so great and these may be preferred on environmental grounds.

Environment The underground cable is invisible. However there can be no building over it or large trees planted since in the event of a fault it must be possible to dig a suitable hole to effect a repair. The heat produced by an e.h.v. cable can affect the soil around it thus modifying the plant growth in the immediate vicinity.

The overhead line has conductors and supports which are sometimes visible for long distances. Electrical discharges from the lines can cause radio interference.

Reliability There is little difference in the reliabilities of the two systems. The overhead line can be struck by lightning whereas the underground cable is at the mercy of earth moving machinery especially when roads are remade or trenches for other services are dug. Occasionally a cable will develop a small hole due to movement over a stone for example giving rise to water ingress followed by an explosion but this is thankfully rare.

Fault finding

Overhead lines are patrolled regularly on foot or by helicopter. Broken insulators can be seen and by using infra-red detection equipment local hot spots can be found possibly in compression joints where two lengths of conductor have been joined. Repairs are reasonably cheap since the line can be taken down, insulators replaced, joints remade and towers repainted at almost any time. If an underground cable develops a fault electrical methods have to be used to locate it. Unless the route is precisely known and the test accurately carried out a great deal of digging is required before the fault can be found. When it has been located the repair is expensive especially on e.h.v. cables.

SWITCHGEAR, DEFINITIONS AND USES

Circuit breaker. A circuit breaker is a mechanical device for making and breaking an electrical circuit under *all* conditions.

Switch. A switch is a device for making and breaking a circuit which is carrying a current not greatly in excess of normal loading.

Isolator. An isolator is a means of isolating or making dead a circuit which is not carrying current at the time (like pulling out a fuse in the home so that work may be carried out in safety on a circuit). The isolator may be used to close a circuit on to load.

The Electricity Supply Regulations state that no piece of electrical equipment may be connected to the mains unless the circuit incorporates a device which will disconnect that equipment automatically in the event of a fault.

According to the definitions above, a circuit breaker is such a device. These are made in miniature form for domestic use with current ratings of between 5 A and 60 A at 240 V while there are larger sizes for industry and transmission and distribution substations which can deal with the highest voltages and currents presently in use.

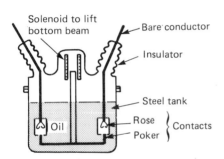

Figure 1.5 Bulk oil circuit breaker

A fuse is often used in place of a circuit breaker in circuits operating up to 11 kV but once it has operated to clear a fault it has to be replaced. This takes time and the larger sizes are very expensive. A circuit breaker can be reclosed after clearing a fault and in addition it may be useful in the rôle of a switch, making and breaking circuits under normal conditions.

Switchgear may be of either the indoor or outdoor variety. For use indoors all electrical conductors are completely enclosed. For use outdoors the circuit breakers are made completely weatherproof. The circuit conductors and isolators are enclosed in 11 kV to 415/240 V substations but at higher voltages they are bare metal insulated from earth using porcelain or glass insulators as described for overhead lines.

Figure 1.5 shows a bulk oil circuit breaker for use outdoors. It is suitable for use up to 132 kV. *Figure 1.6* shows an air circuit breaker employed up to about 11 kV. It may be used indoors or outdoors according to the type of enclosure.

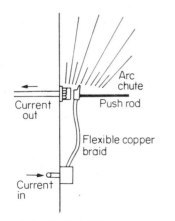

Figure 1.6 Air circuit breaker

SUBSTATION LAYOUT AND EQUIPMENT

The small substation shown in *Figure 1.7* is typical of the thousands of such installations feeding factories, housing and commercial premises. It is connected to the 11 kV ring main shown in *Figure 1.1*.

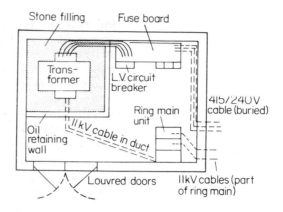

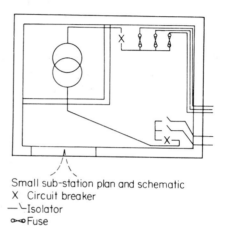

Small sub-station plan and schematic
X   Circuit breaker
—└Isolator
o–oFuse

Figure 1.7

Considering *Figure 1.7*; the ring main is equipped with two isolators at each tap off point and these are normally closed making the 11 kV ring complete. The transformer is fed through a circuit breaker. If a fault occurs on the transformer itself the circuit breaker should open. If a fault occurs on the low voltage system it should be cleared by the low voltage circuit or one of the fuses. Should a fault occur and the correct clearance not take place the circuit breakers controlling the 11 kV ring main (not shown) would operate so making the whole ring dead, depriving other consumers of their supply.

Circuit breakers may be of the oil immersed or air break type. Switches would be used to control individual low voltage circuits for lighting and heating and an automatic switch called a contactor would be used for motor control.

The transformer in the substation will be rated at 11 000 to 433 V nominally. On the high voltage side of the transformer there will be a tap changer. This is a manually operated device which is adjusted off load to alter the number of turns on the winding and has the effect of altering the output voltage to the desired level. The transformer rating will be 300, 500, 750 or 1000 kW. It will generally be oil filled using special transformer mineral oil and the heat produced will be carried by the oil to cooling tubes on the outside of the transformer whence it is convected away by natural air circulation. For this reason the building must be adequately ventilated. This is achieved by fitting louvres in the door of the substation and air bricks or more louvres high up in the walls. Very large transformers are fan cooled but these are installed outdoors. The transformer is either mounted on a plinth in the centre of a large pit filled with stones or there is a small wall built around it. If the transformer suffers a fault and oil leaks out it must be contained to minimise the risk of fire spreading.

In a very large substation with many circuit breakers fire resistant barriers are used to create small sections and automatic fire fighting equipment is fitted in each section using either carbon dioxide gas or fine water spray. Should a fire occur on one circuit breaker only the equipment in the affected section operates.

THE THREE-PHASE SYSTEM

In the three-phase generator the magnetic field on its rotor links in sequence with three equally spaced windings or phases on its stator inducing sinusoidal voltages with equal maximum values in each of them.

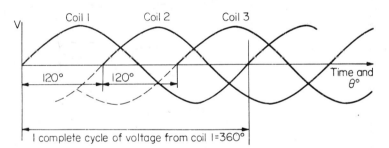

Figure 1.8

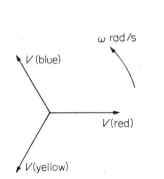

Figure 1.9 Phase voltage phasors in a 3-phase system

These are often known as the Red, Yellow and Blue voltages respectively. These are shown in *Figures 1.8* and *1.9*. The phasors rotate anticlockwise and the red phase voltage wave is followed by that of the yellow phase and again by that of the blue phase.

Figure 1.10 shows a schematic diagram of a generator, its windings physically displaced from each other by 120° and with one end of each winding connected to earth. Because of its appearance this is known as the star connection. The supply lines are labelled red, yellow and blue respectively and the wire connected to the common point is known as the neutral. The voltage from any output line to the neutral is called the phase voltage and the voltage between any pair of supply lines is called the line voltage.

The line voltage = $\sqrt{3}$ × the phase voltage

For example where the phase voltage is 240 V, the line voltage is $\sqrt{3} \times 240 = 415$ V. Single phase loads are connected between any line and the neutral wire and this is the normal situation in the home. One house in a street is connected between the red line and neutral,

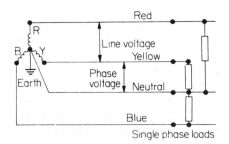

Figure 1.10

the next house between the yellow line and neutral and the next from blue to neutral.

In industry three-phase loads may use three or four wires.

Load balancing

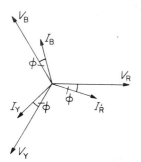

Figure 1.11

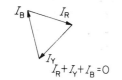

Figure 1.12

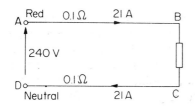

Figure 1.13

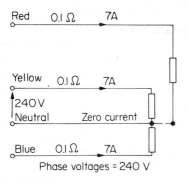

Figure 1.14

Consider now the case of three single-phase loads being equal in magnitude and phase.

Figure 1.11 shows the three phase voltages as in *Figure 1.9* together with the current phasors. These are all equal in magnitude and lag on the phase voltages by an angle $\phi°$. All these currents flow from their respective lines to the neutral. The current in the neutral wire must therefore be the sum of these three currents.

Figure 1.12 shows the phasor addition so that the neutral current proves to be zero. There is no current flowing in the neutral wire back to the supply generator when the loads are balanced.

It is desirable that over the country as a whole the loads shall be balanced over the three phases since this minimises the cable losses and voltage drops in the supply lines. In addition, there are problems in the generators themselves when the three phases are not equally loaded.

Let us consider the power loss and voltage drops in an unbalanced system.

The worst out of balance is where all the load is on a single phase.

Figure 1.13 shows a single phase resistive load drawing a current of 21 A from a 240 V supply along a cable in which line and neutral resistances are both 0.1 Ω.

The power loss = I^2R watts

In the line, power loss = $21^2 \times 0.1 = 44.1$ W. This is repeated in the neutral wire so that the total loss = 88.2 W.

The fall of potential from A to B = $IR = 21 \times 0.1 = 2.1$ V
The fall of potential from C to D = $IR = 21 \times 0.1 = 2.1$ V

The potential difference from B to C = $240 - 2.1 - 2.1 = 235.8$ V

Now consider the effect of obtaining the same power (closely) by using three separate single-phase resistive loads each drawing 7 A spread out over the three phases. The phase voltages are 240 V as before. There is no neutral current since the loads are balanced.

The power loss in each line = $I^2R = 7^2 \times 0.1 = 4.9$ W
In three lines the power loss = $3 \times 4.9 = 14.7$ W
The fall of potential along each line = $IR = 7 \times 0.1 = 0.7$ V
The potential difference across each load = $240 - 0.7 = 239.3$ V

Balancing the loads over the three phases causes the loss in the neutral wire to be eliminated whilst reducing the voltage drops in the cable.

INDUSTRIAL INSTALLATIONS

General arrangement

A typical substation for a factory is shown in *Figure 1.7*. Such a substation would most likely be situated on the factory premises.

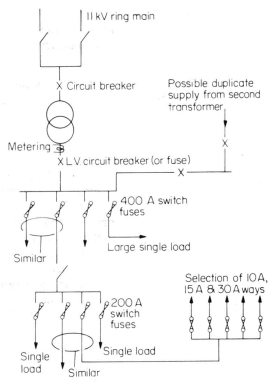

Figure 1.15 Small factory schematic

Metering of energy consumed and the maximum demand made by the factory on the supply system is carried out using current transformers fitted in, or adjacent to, the main low voltage circuit breaker or fuses. The current transformers feed a kilowatt-hour meter which at 415/240 V derives its voltage coil supply directly from the bus bars without the use of potential transformers. The main distribution fuse board is fitted in the substation.

In *Figure 1.15* this is shown as being equipped with switches incorporating fuses rated at 400 A. The actual size employed will depend on the rating of the equipment installed in the factory.

Some large single loads may be fed directly from this point, the actual control being by local switch or contactor. There will be several ways feeding further fuse boards in particular workshops. These will feed local heavy loads or further fuse boards for lighting, heating, small tools and processes.

In *Figure 1.15* what is known as single-line representation has been used: instead of drawing three conductors representing the three phases throughout the diagram, a single line and a single fuse are used to represent a three-phase system. This means, for example, that a single 400 A switch fuse contains in fact three fuses and when it operates, all three phases are broken. The fuse board shown, with the 10 A, 15 A and 30 A ways, will have these spread, as far as possible, over the three phases.

On single-phase systems, although there will be two wires for each circuit, often only one line is drawn representing a single cable. In tables of resistance and volt drops these will be quoted as a 'loop' value, that is, the values for both go and return conductors are lumped together.

DISTRIBUTION SYSTEMS

Radial system

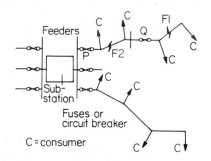

Figure 1.16

A radial system for the distribution of electrical energy is shown in *Figure 1.16*. A substation supplies consumers C through radial distributors which fan out from the substation. A fuse or circuit breaker protects each distributor. Some distributors have subsidiary fuses which are of lower rating than the main fuse somewhere along their length. In the event of a fault on the feeder the relevant fuse clears leaving all consumers on that section without a supply. In *Figure 1.16* a fault F1 would cause fuse Q to clear leaving two consumers without a supply. A fault at F2 would cause fuse P to clear leaving all four consumers without a supply. Since there is no alternative method of supply to these consumers repairs to the line have to be carried out before they can be reconnected.

The consumer on the end of each distributor suffers voltage reductions as the load on that distributor increases and to minimise this the cross-sectional area of the conductors is large. This is therefore an expensive system to install and offers poor security of supply.

Fault finding is relatively simple since tests need only be done on the cleared section of line.

Ring system

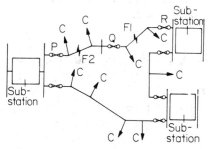

Figure 1.17

Addition of further substations feeding the other ends of the radial feeders as shown in *Figure 1.17* effectively converts the system into a ring. This substantially reduces the voltage drops along the distributors and enables savings in conductor cross-sectional areas and costs to be made. A fault F1 would result in fuses Q and R clearing leaving two consumers without a supply. A fault F2 would result in fuses P and Q clearing and again only two consumers would lose their supply instead of four with the radial system.

At 11 kV a ring system employing isolators at each load point enables greater security of supply to be achieved. In *Figure 1.18* a fault F1 can be cleared by opening isolators I_2 and I_3 and no other section need be disconnected provided that a second fault does not occur. Meanwhile repairs can be carried out.

Interconnection of two points on the ring makes the system more versatile while reducing voltage drops and cable losses. This makes for greater expense however.

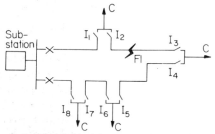

Figure 1.18

If circuit breakers are used instead of isolators, overcurrent relays which are sensitive to the direction of current flow can be used to automatically isolate a faulty section. When isolators are used there will be an interruption of supply to consumers while the fault is located and the isolators are manually operated.

Reduction of voltage drops and cable losses together with increased security of supply are achieved by increasing the number of substations and the degree of interconnection. However, the expense rapidly increases with complexity and the actual arrangement employed is the best which can be obtained at a realistic cost to the consumer.

DISTRIBUTOR CALCULATIONS

Radial

In *Figure 1.13* we saw a single resistive load being fed from a 240 V supply. Using a single line representation, the diagram can be redrawn as shown in *Figure 1.19*. The resistance marked is that of both go and return conductors.

Figure 1.19

The load voltage = $240 - 21 \times 0.2$
$$= 240 - 4.2$$
$$= 235.8 \text{ V as previously.}$$

The method is suitable for both d.c. and single phase a.c. where resistance only is considered. When the loads or lines have inductance or capacitance there are phase angles to consider and arithmetic addition and subtraction of voltages gives incorrect results.

Let us now consider feeder with three loads.

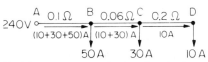

Resistance are go and return (loop) values

Figure 1.20

Example 1. The details of a radial feeder are shown in *Figure 1.20.*
Calculate: (a) The load voltages
            (b) The power lost in the cable
            (c) The power developed by each load
            (d) The efficiency of the system.
Applying Kirchhoff's first law to each load point:
The current in section CD of the feeder = 10 A
Section BC carries load currents C and D = 30 + 10 = 40 A
Section AB carries the total load currents = 40 + 50 = 90 A
The voltage drop between A and B = $I_{AB} \times R_{AB} = 90 \times 0.1$
= 9 V
Voltage at load B = 240 − 9 = 231 V
Power loss in section AB = $(I_{AB})^2 R_{AB} = 90^2 \times 0.1 = 810$ W
Power developed by load B = $V_B I_B = 231 \times 50 = 11\,550$ W
Repeat for section BC
Voltage drop from B to C = $40 \times 0.06 = 2.4$ V
Voltage at load C = 231 − 2.4 = 228.6 V
Power loss in section BC = $40^2 \times 0.06 = 96$ W
Power developed by load C = $228.6 \times 30 = 6858$ W

Repeat for section CD
Voltage drop from C to D = $10 \times 0.2 = 2$ V
Voltage at load D = $228.6 - 2 = 226.6$ V
Power loss in section CD = $10^2 \times 0.2 = 20$ W
Power in load D = $226.6 \times 10 = 2266$ W
Total load powers = $2266 + 6858 + 11\,550 = 20\,674$ W
Total losses = $20 + 96 + 810 = 926$ W

$$\text{Efficiency} = \frac{\text{Power in loads}}{\text{Total power input}} = \frac{20\,674}{20\,674 + 926} \text{ OR } \frac{20\,674}{240 \times 90}$$
$$= 0.957 \text{ p.u.}$$

Distributor fed at both ends   Now consider the effect of reinforcing the system by feeding point D either from another substation or from the same substation through an extra length of cable or line.

Example 2.   Re-calculate (a) to (d) in Example 1 with the supply reinforced as shown in *Figure 1.21*.
In this case we firstly have to determine how much current is supplied from each end of the feeder.
Consider a current I_1 to be entering from the left hand end.
The current in the section AB = I_1A and the volt drop between A and B = $I_1 \times 0.1$ V.
At point B a load current of 50 A is supplied so that the current flowing on in section BC must be $(I_1 - 50)$ A and the volt drop from B to C = $(I_1 - 50) \times 0.06$ V.
Similarly, the current in section CD = $(I_1 - 50) - 30$
= $(I_1 - 80)$ A and the volt drop from C to
D = $(I_1 - 80) \times 0.2$ V.
In section DE the current is $(I_1 - 90)$ A and the volt drop from D to E = $(I_1 - 90) \times 0.1$ V.
The voltage at A − all the volt drops along the line = voltage at E.
This is as in Example 1 except in this case we know the voltage at E = 240 V.
Therefore
$240 - I_1 \times 0.1 - (I_1 - 50) \times 0.06 - (I_1 - 80) \times 0.2$
$- (I_1 - 90) \times 0.1 = 240$
Multiply out the brackets
$240 - 0.1I_1 - 0.06I_1 + 3 - 0.2I_1 + 16 - 0.1I_1 + 9 = 240$

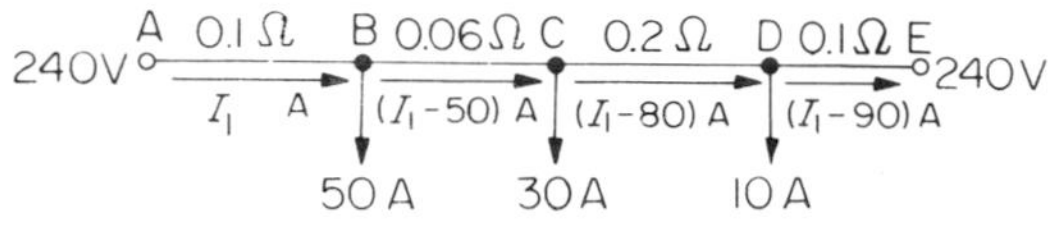

Figure 1.21

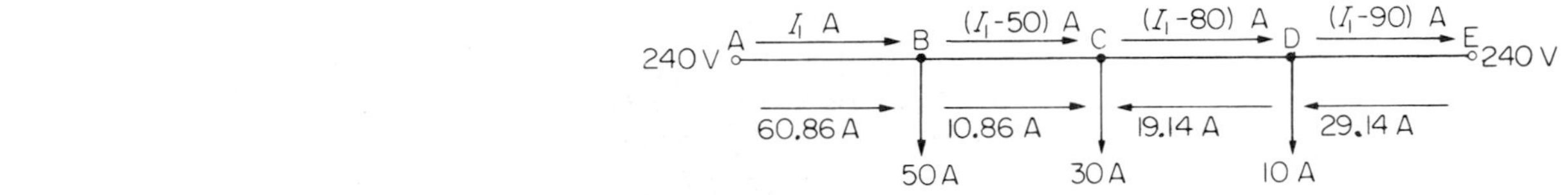

Figure 1.22

Transpose
$$240 - 240 + 3 + 16 + 9 = 0.1I_1 + 0.06I_1 + 0.2I_1 + 0.1I_1$$
$$28 = 0.46I_1$$
$$I_1 = 60.86 \text{ A}$$
Redraw the diagram as in *Figure 1.22*.
In section BC we have $(60.86 - 50)$ A = 10.86 A
In section CD we have $(60.86 - 80)$ A = −19.14 A. The negative sign indicates a reversal of current direction from that shown in *Figure 1.21* and this is seen to be logical since the 30 A load receives 10.86 A from end A and 19.14 A from end E.
Voltage at B = $240 - 60.86 \times 0.1 = 233.9$ V.
Power loss in section AB = $60.86^2 \times 0.1 = 370.4$ W.
Power in the load B = $233.9 \times 50 = 11\,695$ W.
Voltage at C = $233.9 - 10.86 \times 0.06 = 233.25$ V.
Power loss in section BC = $10.86^2 \times 0.06 = 7.08$ W.
Power in load C = $233.25 \times 30 = 6997.5$ W.
Voltage at point D. Since current flows from D to C, the voltage at D must be greater than that at C. The minimum voltage on the distributor is at load C and currents flow from both ends to this point.
Voltage at D = $233.25 + 19.14 \times 0.2 = 237.08$ V.
Power loss in section CD = $19.14^2 \times 0.2 = 73.26$ W.
Power in load D = $237.08 \times 10 = 2370.8$ W.
We know that the voltage at point E is 240 V. Let us check the calculations by adding the volt drop from E to D to the voltage at D
$$237.08 + 29.14 \times 0.1 = 240 \text{ V}$$
Power loss in section DE = $29.14^2 \times 0.1 = 84.91$ W.
Total load powers = $11\,695 + 6997.5 + 2370.8 = 21\,063.3$ W.
Total losses = $370.4 + 7.08 + 73.26 + 84.91 = 535.65$ W.

$$\text{Efficiency} = \frac{21\,063.3}{21\,063.3 + 535.65} = 0.975 \text{ p.u.}$$

Notice that by feeding the system at both ends each of the load voltages has been increased. The cable losses have been reduced so increasing the efficiency.

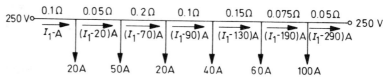

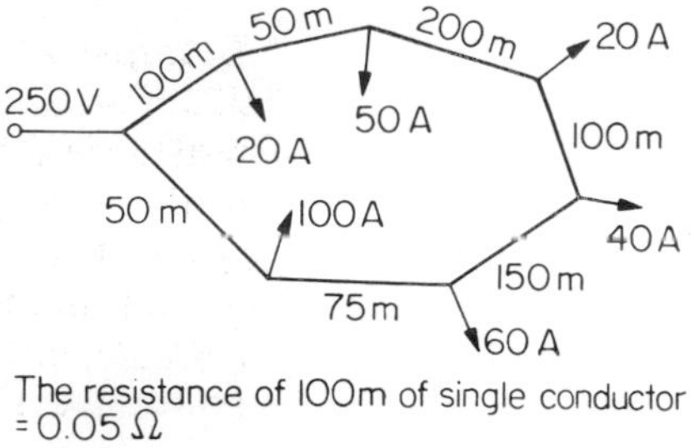
250 V
100m
50 m
200 m
20 A
50 A
20 A
100 m
50 m
100 A
40 A
150 m
75 m
60 A
The resistance of 100m of single conductor
= 0.05 Ω

Figure 1.23

Example 3. For the ring main shown in *Figure 1.23*, determine the current in each section and the minimum load voltage.

Redraw the diagram putting in the go and return resistances. A 100 m go and return has a resistance of 2 × 0.05 = 0.1 Ω. A ring may be considered as a feeder fed at both ends at the same voltage.

250 V
0.1Ω
0.05Ω
0.2Ω
0.1Ω
0.15Ω
0.075Ω
0.05Ω
250 V
I_1-A
(I_1-20)A
(I_1-70)A
(I_1-90)A
(I_1-130)A
(I_1-190)A
(I_1-290)A
20A
50A
20A
40A
60A
100A

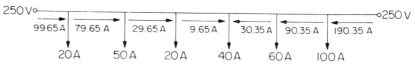

Figure 1.24

Starting with a current I_1 flowing towards the 20 A load.

$250 - 0.1I_1 - 0.05(I_1 - 20) - 0.2(I_1 - 70) - 0.1(I_1 - 90)$
$- 0.15(I_1 - 130) - 0.075(I_1 - 190) - 0.05(I_1 - 290) = 250$

$250 - 0.1I_1 - 0.05I_1 + 1 - 0.2I_1 + 14 - 0.1I_1 + 9 - 0.15I_1$
$+ 19.5 - 0.075I_1 + 14.25 - 0.05I_1 + 14.5 = 250$

$72.25 = 0.725I_1$

$I_1 = 99.65$ A

250 V
250 V
99.65 A
79.65 A
29.65 A
9.65 A
30.35 A
90.35 A
190.35 A
20A
50A
20A
40A
60A
100A

Figure 1.25

The minimum voltage occurs at the 40 A load. The voltage at this point may be determined starting at either end.

From the left

$250 - 99.65 \times 0.1 - 79.65 \times 0.05 - 29.65 \times 0.2 - 9.65 \times 0.1$
$= 229.16$ V

From the right

$250 - 190.35 \times 0.05 - 90.35 \times 0.075 - 30.35 \times 0.15$
$= 229.16$ V

Example 4. A two-wire feeder is 275 m in length. It is fed at 238 V at one end and 245 V at the other end. The feeder loop resistance (go and return conductors) is 0.05 Ω/100 m. Three loads of 50 A, 100 A, and 25 A are being fed at distances of 100 m, 150 m, and 200 m respectively from the end of the feeder being supplied at 238 V.

Calculate (a) the value of the minimum load terminal voltage on the system and (b) the overall efficiency of the system.

First draw the system and, using the distances, calculate and mark on the feeder the loop resistances (*Figure 1.26*).
For example, 75 m of feeder will have a resistance of 75/100 × 0.05 = 0.0375 Ω.

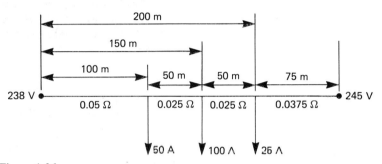

Figure 1.26

Let I_1 A enter at the left-hand end. The currents will be as shown in *Figure 1.27*.

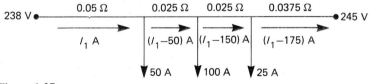

Figure 1.27

238 − (all the volt drops along the feeder) = voltage at the other end = 245 V

$238 - 0.05I_1 - 0.025(I_1 - 50) - 0.025(I_1 - 150)$
$\quad - 0.0375(I_1 - 175) = 245$

$238 - 0.05I_1 - 0.025I_1 + 1.25 - 0.25I_1 + 3.75 - 0.0375I_1 + 6.56$
$\quad = 245$

$238 + 1.25 + 3.75 + 6.56 - 245 = (0.05 + 0.025 + 0.025$
$\quad + 0.0375)I_1$

$4.56 = 0.1375I_1$

$I_1 = 33.2\ \text{A}$

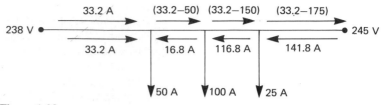

Figure 1.28

The minimum terminal voltage is at the 50 A load, since currents flow towards this load from both ends (*Figure 1.28*).

Load voltage = 238 − 33.2 × 0.05

= 236.34 V   (Answer (a))

The same result is achieved by working from the other end of the feeder, subtracting the volt drops from 245 V.

(b) Losses = I^2R watts. Considering each section in turn:

Losses = 33.2² × 0.05   =   55.1 W

16.8² × 0.025   =   7.06 W

116.8² × 0.025   = 341.06 W

141.8² × 0.0375 = 754.02 W   Total losses = 1157.2 W

Total power into the feeder = 33.2 A at a potential difference of 238 V = 33.2 × 238

= 7901.6 watts

and 141.8 A at a potential difference of 245 V = 245 × 141.8

= 34 741 watts

Total power input = 42 642.6 W

$$\text{Efficiency} = \frac{\text{output}}{\text{input}} = \frac{\text{input} - \text{losses}}{\text{input}}$$

$$= \frac{42\,642.6 - 1157.2}{42\,642.6}$$

$$= 0.973 \text{ p.u. } (97.3\%)$$

PROBLEMS FOR SECTION 1

5 Calculate the value of current necessary to transmit a power of 100 kW at a power factor of 0.7 lagging at (i) 500 V (ii) 5000 V. If in each case the conductor resistance is 0.2 Ω, calculate the two power losses.

6 What is the function of the 400 kV system in the UK?

7 What is the function of the 132 kV and 33 kV systems in the UK?

8 What type of insulation is used in e.h.v. underground cables?

9 Why does the cost of transmission equipment rise as the voltage levels are raised?

10 In the UK, domestic premises are supplied at between 230 V and 250 V. In some other countries 110 V systems are used. What are the advantages and disadvantages of such a system?

11 What are the advantages of using steel lattice towers to support overhead power lines as compared with wooden poles?

12 List the advantages and disadvantages of an overhead transmission system as compared with one using underground cables.

13 What is the function of a circuit breaker?

14 What difference in design would you expect to find in an indoor circuit breaker as compared with one situated outdoors?

15 What is the function of the stone-filled pit beneath an oil-filled transformer?

16 What is the function of a tap changer?

17 Why are large indoor switching and transforming stations sectioned?

18 The line voltage of a synchronous three-phase alternator is 33 kV. What is the value of the voltage from one line to the neutral point?

19 The output voltage from a three-phase transformer is 125 V per phase. What is the value of the line voltage?

20 Why is care taken to balance loads across the phases of a three-phase system?

21 A radial feeder *ABCD* is fed at *A* at 200 V. The loads are 20 A, 10 A and 10 A at *B*, *C* and *D* respectively. The loop resistances of the cables are:
$AB = 0.1\,\Omega$, $BC = 0.15\,\Omega$, $CD = 0.05\,\Omega$.
Calculate the efficiency of the system under these conditions.

22 A feeder *ABCD* is fed at *A* and *D* at 220 V. A load of 20 A is situated at *B* which is 100 m from *A*. A load of 30 A is situated at *C* which is 120 m from *D*. The feeder is 420 m in length. The resistance of 100 m of *single* conductor is $0.025\,\Omega$.
Determine the currents in each section of the feeder and the minimum voltage.

23 For the feeder shown in *Figure 1.29* determine the power in the load which has the minimum potential difference across it.

24 A ring main *ABCDEFGA* is fed at *A* at 250 V.
$AB = 50$ m, $BC = 50$ m, $CD = 100$ m, $DE = 75$ m, $EF = 75$ m, $FG = 150$ m and $GA = 100$ m.
A 100 m loop (go and return) of the ring has a resistance of $0.1\,\Omega$. The loads are as follows:
$B = 20$ A, $C = 30$ A, $D = 10$ A, $E = 50$ A, $F = 20$ A, $G = 25$ A.
Determine the values of the currents in each section of the ring and the value of the minimum potential difference at a load.

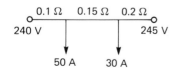

Figure 1.29

25 For the feeder shown in *Figure 1.30*, calculate (i) the individual load voltages and powers and (ii) the efficiency of the system. A 100 m loop of the feeder has a resistance of 0.06 Ω.

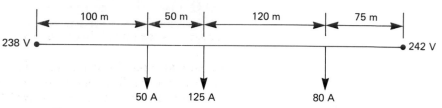

Figure 1.30

2 Regulations

Aims: At the end of this chapter you should be able to:

Explain the need for, and discuss the scope of the Electricity Supply Regulations, the Regulations for Electrical Installations, the Health and Safety at Work Act and the Electricity at Work Regulations.

SAFETY AND REGULATIONS

Inadequate control of electricity can give rise to serious dangers due to fire and shock. In the early days of the electricity public supply system the powers involved were small and faults were frequent, often resulting in whole plants being shut down. At least one power station was destroyed by fire when a fault developed and was not cleared. With the vast amounts of power involved today such a situation cannot be allowed to develop and over the years regulations applying to suppliers and consumers have been drawn up to prevent dangerous situations and accidents as far as is reasonably practicable. A major accident causes the regulations to be reviewed to see how best to prevent a future similar occurrence but it must be said that many of the original regulations were so well and widely cast that remarkably few changes have been required over the years.

These regulations are:

The Electricity Supply Regulations 1988;
The Institution of Electrical Engineers' Regulations for Electrical Installations.

In addition all workers are covered by *The Health and Safety at Work Act.* Under this Act *The Electricity at Work Regulations 1989* have been brought into force.

There are also special regulations applying to coal mines, quarries, cinemas, places of public entertainment and oil refineries.

THE ELECTRICITY SUPPLY REGULATIONS

These are administered by the Engineering Inspectorate of the Electricity Division of the Department of Energy. They are designed to secure the safety of the public and to ensure a proper and safe supply of electrical energy. They impose requirements on the suppliers of electrical energy and apply to the electrical distribution system covering construction, operation, protection and earthing, maintenance and safety up to the consumers' terminals.

These regulations may apply additionally to those imposed by the Electricity at Work Regulations.

Certain sections relate to the consumer's installation in as much as they give the Supply Companies powers to insist upon certain standards of work before the supply can be connected. The standards generally required are those laid down in the IEE Regulations for Electrical Installations. In addition, a supply may be refused to a consumer whose equipment is likely to affect other consumers adversely.

The Supply Companies have to declare the voltage of the supply and maintain this within 6 per cent of that value.

The Generating Companies maintain the frequency within specified limits and maintain an average of 50 cycles per second (50 hertz) over 24 hours (so keeping electric clocks correct).

Both Generating and Supply Companies have very strict sets of safety rules and operate a 'Permit to Work' system which ensures an extremely low accident rate.

THE INSTITUTION OF ELECTRICAL ENGINEERS' REGULATIONS FOR ELECTRICAL INSTALLATIONS

The overall objective of these regulations is to ensure the protection of people and livestock from fire, shock or burns from any installation that complies with their requirements.

The current edition of these regulations is the sixteenth. This was published in May 1991 and came into force on 1 January 1993. Since October 1992 the IEE Regulations have become a British Standard: BS 7671 1992 Requirements for Electrical Installations.

The regulations are based on the international regulations drawn up by the International Electrotechnical Commission with the aim of having a common set of regulations for all EC countries. (Worldwide consensus on electrical safety is a desirable goal but probably not yet practicable.)

The 15th edition regulations were complete in themselves and included much guidance material. The 16th edition omits the guidance material: only the legal skeleton is present. There are to be eight sets of guidance notes to help with their interpretation and aid good practice. These will cover broadly the material in the guidance notes of the 15th edition.

The regulations covering a particular aspect of working, for example earthing or circuit protection, are almost invariable subdivided and are to be found in several different parts of the Regulations. The overall picture can only be obtained after a consideration of all the relevant sections, which involves much use of the index. For this reason, only a very limited number of topics are quoted here with a selection of locations where references may be found.

- Electrical conductors shall be of sufficient size and current-carrying capacity for their intended purpose. [434–03–03]
  (The location quoted breaks down into: Part 4, Chapter 3, Section 4, Group or Subsection 3, the 3rd Regulation.)

- All conductors shall be suitably insulated or protected to prevent danger, bearing in mind the environment in which they are working. [522–01 to 522–05 and parts of 523 and 527]

- Circuits shall be protected against overcurrents. [432–02 432–03 and parts of 433 and 473]

- Circuits must be adequately protected against fault currents. [432–02, 432–04 and elsewhere]

- The requirements for equipment to be adequately earthed and the methods appear in Sections 130–04, 542–04 and others.

- The requirements for installation of lighting circuits appear in Sections 422–01, 553–03 and others.

In Part 4, tables of cable sizes for different currents are provided together with factors to be applied where the cables operate in a high ambient temperature or where several cables are grouped together, possibly within a conduit. Different factors have to be used according to whether the circuit is protected by an HRC fuse, a miniature circuit breaker or a rewireable fuse. Where a rewireable fuse is used, allowance must be made for the fact that it is able to carry a large overcurrent for a considerable time before clearing. For this reason no rewireable fuse can have a rating greater than 0.725 times that of the smallest cable it is protecting. [433–02]

Compliance with the IEE Regulations is implicitly required by The Electricity at Work Regulations (1989), so that, should an accident occur due to non-compliance with these regulations, legal proceedings could result.

Certain installations where there are exceptional risks must also comply with additional Statutory Regulations, namely those in cinemas, coal mines, quarries, oil refineries and horticultural and agricultural installations. Work conforming to the IEE Regulations will generally satisfy these Statutory Regulations in so far as they relate specifically to the associated buildings. Installations of a special character will require the additional advice of a suitably qualified specialist. The Regulations recognise that not all eventualities can be covered.

THE HEALTH AND SAFETY AT WORK ACT

The Health and Safety at Work Act (HASAWA) came into force in 1974. It was introduced as a result of a report by the Robens Committee on Health and Safety, one of the findings of which was that a considerable number of accidents were the result of apathy on the part of workers and management alike.

Up to this time there had been a number of Factories Acts, the first being in 1833 which limited the working hours of children. All these Acts were aimed at premises, e.g. offices, workshops and railway premises. They aimed at making these safer places to work in. The Health and Safety at Work Act, however, is directed at people and activities. The underlying objective is to involve everybody at the workplace, whether this be in a factory or on a farm, at a bench or driving a lorry, and to create an awareness of the importance of achieving high standards of health and safety. This gives far greater coverage than the old Factories Acts, protecting everybody at work (except domestic servants employed in private households). It also extends the protection to the general public and to visitors to places of activity who are not employees.

The Health and Safety at Work Act is an 'Enabling Act'. This means that it does not itself contain a lot of detailed regulations but enables other, often older acts, such as The Offices, Shops and Railway Premises Act and the more recent Electricity at Work Regulations 1989, to be used to cover most working situations.

Examples will serve to demonstrate the point. Previous to the HASAWA there was no specific Act for schools and colleges, so that, with a few exceptions, the staff, pupils and students were not covered by legislation. The school secretary, recognised as an office worker, was protected under the Offices, Shops and Railway Premises Act and could legally demand that the office temperature be maintained at a satisfactory level, whereas teaching colleagues in an adjacent room were not so covered. In a college laboratory a pressure vessel or boiler not originally covered by legislation would now be equated to its equivalent situated in a factory and would therefore need to be inspected and tested in the same way.

Outline of the requirements of the HASAWA

1 The employer must ensure, as far as is reasonably practicable, the health, safety and welfare of all his employees by the provision of:
 (i) safe plant and equipment
 (ii) safe materials handling, storage and transport facilities
(iii) safe systems of work, with adequate training and supervision
(iv) safe premises and working environment.

In certain circumstances he must provide for periodical medical examination of his employees. This will be particularly relevant where toxic substances are being handled, for example.

He must ensure that no process in the workplace yields harmful emissions which could damage the health of employees or the general public.

A copy of the firm's safety policy must be available to all employees. This lays down the procedures for safe working, and who is responsible for carrying out the policy. This person must be someone with executive power.

2 Employees must take reasonable care of their own health. They must, where relevant:
 (i) wear safety glasses, ear muffs and protective clothing
(ii) cooperate with colleagues and supervisors to promote safe working.

The situation where a machine guard has been removed in order to clear an obstruction and not subsequently replaced, or removed to facilitate extra production, might be envisaged. The person removing the guard, any person using the machine in an unguarded state and the supervisor who should see that the guards are maintained in a working condition could all be deemed negligent.

Manufacturers, suppliers and importers have to ensure that goods and equipment supplied are safe for use in the manner for which they are designed. This responsibility can be traced back to the designer of the equipment if necessary. Equipment manufactured and tested to the levels set by the relevant British Standard will satisfy this requirement in every way. If the material is imported with no British Standard number it must have a 'Harmonisation Document' which will guarantee that it has been designed and manufactured to a standard at least as good as the British Standard.

All substances must be adequately labelled to ensure that they will be correctly used and, particularly when they are being transported in bulk, to enable the emergency services to deal quickly and correctly with any spillage.

The Health and Safety Inspectorate

The enforcing body of the HASAWA is the Health and Safety Executive which includes Her Majesty's Factory Inspectorate. This body has overall responsibility for the day-to-day operation of the Act and its enforcement. The factory inspectorate is now divided into Area Offices and further into Industry Groups to provide expert inspectors for particular industries. However, the scope of the Act is much wider than industry, and under the 'Enforcing Authority Regulations' local authorities are responsible for its enforcement in certain areas of its application. For instance, the County or Town Councils oversee offices to see that they comply with the Offices, Shops and Railway Premises Act; the Local Education Authorities oversee schools and colleges of further education.

The inspectors have wide powers. These include the right to inspect and investigate anything in the course of their duties. They may enter premises at any 'reasonable time', or at any time at all if they have reason to suppose that a dangerous situation exists within. If the inspector thinks that his entry may be impeded, he may bring a police officer with him. If the situation warrants it he may issue either an 'Improvement Notice' or a 'Prohibition Notice'. The Improvement Notice will require a situation or process to be improved to decrease the risk of danger. The improvement will have to be carried out within a specified period of time. The recipient of the notice may appeal, but his appeal does not increase the length of time allowed. Failure in the appeal means that the improvement must be carried out by the original date. A Probibition Notice means that the process must stop immediately. It can only restart when the process has been made safe and has been approved by the inspectorate.

THE ELECTRICITY AT WORK REGULATIONS (1989)

This set of regulations has been made under the Health and Safety at Work Act. They have been drawn up to ensure safe working with electricity in the workplace and aim to prevent unsafe installations and practices which otherwise might be employed out of ignorance or to save costs. There are 33 regulations in all: sixteen general regulations are followed by others which relate to mines only. There are distinct similarities between the general regulations and the IEE Regulations. This is not surprising, since safe use of electricity at work and safe installation practice are inevitably closely linked. Since the regulations are designed to promote safe working, it follows that non-compliance will create a hazardous or unsafe condition.

Enforcement of these regulations under the HASAWA is by the Factory Inspectorate.

We will examine some of the general regulations.

Reg 3 Employers, self-employed persons and managers of mines or quarries must comply with the provision of these regulations. In addition it is the duty of every employee to cooperate with his employer in complying. This makes it equally the responsibility of supervisor and supervised to ensure safety. This is reinforcement of Section 7(b) of the HASAWA: every person must take responsibility in association with others for his own safety. In addition, he must not take any action which either directly or indirectly will put anybody else at risk.

Reg 4 All systems must be of such a construction and be maintained in such a condition as to prevent danger. Equipment provided for the protection of persons working on or near electrical equipment must be suitable for that use and maintained in good condition.

Reg 6 Electrical equipment which may be subjected to adverse conditions of weather, heat, chemical attack etc. must be of such a construction as to withstand such conditions.

Reg 8 Any conductor which can become charged as a result of the method of usage or of a fault must have a suitable means of preventing danger to life by contact with that conductor. This will often be by means of earthing (the metal case of a motor, for example).

Reg 10 Every joint or connection must be mechanically and electrically suitable for the situation. This means, for example, that when making an earth connection, the wire should be held mechanically, perhaps by a screw or twisted joint rather than just attached by soft solder which might fail under mechanical movement.

Reg 11 Efficient means must be proved for protecting a system from excess current. Often this is a fuse or circuit breaker actuated by an excess current detection device.

Reg 12 Suitable identification of circuits must be provided together with means of isolating them from the electricity supply.

Reg 13 When equipment has been made dead for work to be carried out, adequate precautions must be taken to ensure that the equipment cannot be re-energised during that work.

Reg 15 Adequate working space and means of access must be provided to safely carry out work on equipment.

Reg 16 Every person carrying out work on equipment must be in possession of such technical knowledge or experience as is necessary to prevent danger or be under the direct supervision of a person so qualified.

COMPARISON BETWEEN THE SCOPES OF THE REGULATIONS

The Supply Regulations apply up to the supply point in the factory or premises. The IEE Regulations relate to installation practice on consumers' premises. The IEE Regulations is a code of practice which is widely accepted in the UK and compliance with them is likely to satisfy the Electricity at Work Regulations 1989 and the Supply Regulations in as far as these affect consumers.

The Electricity at Work Regulations have the force of law behind them. Prosecution can result from non-compliance and the IEE Regulations could be quoted to demonstrate malpractice.

The IEE Regulations are likely to be the basis of a contract between an installation firm and the purchaser of an installation.

PROBLEMS FOR SECTION 2

1   What is the general objective of all electricity supply and wiring regulations?
2   What powers do the Electricity Regulations give to the Supply Companies with respect to consumers' equipment?

3 What do the Electricity Supply Regulations have to say about terminal voltage?

4 What additional safety procedures are adopted by the Generating and Supply companies to minimise accidents?

5 What powers do Factory Inspectors have with regard to unsafe equipment or working practices? How might they initiate corrective measures?

6 What is the difference in scope between the Electricity at Work Regulations and the IEE Regulations?

7 What is the basic objective of the IEE Regulations?

8 What additional regulations may have to be complied with in certain hazardous situations?

9 In a court case involving the Electricity at Work Regulations, what weight might the IEE Regulations have?

10 Which of the regulations/acts mentioned apply to the consumers of electrical energy?

11 Which of the regulations/acts mentioned apply to the suppliers of electrical energy?

12 In what major aspect do the HASAWA and the Electricity at Work Regulations differ from the older Factories Acts?

13 What is a 'Harmonisation Document'?

3 Tariffs and power factor correction

Aims: At the end of this chapter you should be able to:

Define maximum demand, diversity factor, load factor and power factor.
Explain the need for and the use of a two part tariff.
Calculate the charges for electrical energy on various tariffs.
Calculate the savings to an industrial consumer brought about by an improvement in power factor.

GENERATION AND TRANSMISSION COSTS

The cost of generation and transmission of electrical energy is divided into two parts:
1 Capital charges
2 Running charges.

Capital charges

In order to build a power station or a transmission line money is borrowed and interest paid annually. Money is also put aside so that in theory at least, at the end of 25 or 30 years the loan can be repaid. This is called a depreciation allowance and with an expanding system is in fact spent more or less continuously or new plant. Both interest and depreciation charges have to be met whether the plant purchased is used or not.

Running charges

In order to run a power station men must be paid, fuel purchased and repairs carried out. On overhead lines and underground cables repairs must be done and routine testing and inspections carried out. These costs are very nearly proportional to the amount of energy sold. The cost of electrical losses in the system are also included in the running charges.

In the UK electricity is purchased from the Generating Board by the Area Electricity Boards who act as retailers to industry and the general public. The Area Boards and all their consumers must in turn pay both the capital and running charges of the system and therefore all tariffs have a two part structure; a capital charge which is a sum based on the availability of plant to meet the consumers' demand, generators, lines and transformers; and a running charge which is based on the cost of fuel and other resources used in the production of the energy consumed. The running charge is fixed at intervals but the Generating and Area Boards have powers to vary this automatically as fuel costs change by means of a 'Fuel Adjustment Charge'.

For example: cost per kWh = 4.5 p ± 0.000 08 p for every penny by which fuel costs in the region vary from £25/tonne of coal equivalent. Where oil is burnt, the heat output from this and its price are often related to the values for coal to make cost judgements more easy. It might be that oil is more expensive than coal on a particular day but that when burnt it provides more heat, so that the overall cost of energy derived from it could be less. Should the Generating Board decide to reduce prices charged to the Area Boards because of a general fall in fuel costs, the Area Boards may pass this on to their customers as a percentage reduction, e.g. 5 p/kWh − 4% fuel cost reduction.

Very large consumers may be supplied at 132 kV or 33 kV. A supply at these voltages does not involve the use of the rest of the distribution system down to 415/240 V and since less plant is involved there will be a saving in capital charges. In addition there will be a reduction in system losses enabling a slight reduction in running charges to be made.

POWER FACTOR

Most loads on the electricity supply system comprise resistance and inductance in series so that the supply current lags on the voltage by an angle $\phi°$ as shown in *Figure 3.1*.

The amount of this current which is in phase with the voltage and therefore capable of doing work is called the active component of current I_a.

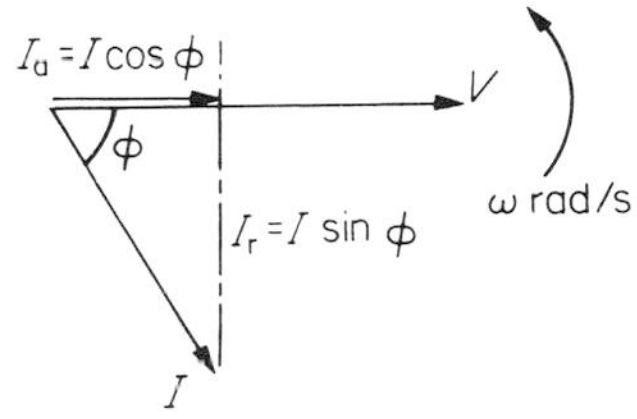

The power in the circuit $= VI_a$ watts
The product $V \times I$ is termed volt-amperes.

$\dfrac{\text{True power}}{VI}$ is the circuit power factor.

Hence power factor $= \dfrac{VI_a}{VI} = \dfrac{I_a}{I}$ which from *Figure 3.1* can be seen to be equal to $\cos \phi$.

Also in *Figure 3.1*, the vertical side of the current triangle I_r can be found since $I_r/I = \sin \phi$ so that $I_r = I \sin \phi$.

In a circuit comprising pure resistance, all the current is in phase with the voltage so that $I = I_a$ and $I_r = 0$. I_r is only present when the circuit has reactance and is therefore known as reactive current.

Taking the current triangle in *Figure 3.1*, multiplying each side by V, the circuit voltage, gives a similar triangle showing the relationship between power, volt-amperes and volt-amperes-reactive as the product $V \times I_r$ is called.

Since the current lags on the voltage in this case these are known as lagging volt-amperes-reactive.

[Notice carefully in *Figure 3.2*: I is the symbol for current whilst A is the unit of current. We write $I = 5$ A for example. Similarly $VI_r = 250\ VA_r$]

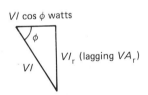

Figure 3.1

Figure 3.2

MAXIMUM DEMAND

The Board must obtain payment from its consumers for the amount of plant involved in its operation irrespective of the number of hours that it runs. Consider a small system of say 100 000 kW capacity, the total capital charges on which are £1.5 million per annum. This represents £15 for each kilowatt of plant capacity per annum. 100 000 consumers each with a 1 kW simultaneous demand should each pay £15 per annum to cover the capital charges. Whether the consumer leaves the equipment on for the whole year or just for the one hour the amount of plant involved is the same and the charge is the same. There would of course be a difference in the energy or running charge between the two cases.

When the Area Boards purchase energy from the Generating Board or industrial consumers from the Area Boards the amount of plant involved is determined by measuring maximum demand. This may be on a kW or kVA basis. In the former case a kilowatt-hour meter has its advance measured during each half hour of the year. A clock times half hours from the hour to half past and from half past to the hour. The number of kWh used in one half hour multiplied by two gives the hourly rate in kWh per hour. [kWh/h = kW]

The largest value of the kW demand in a given period (month, quarter or year) is the maximum demand for the period. Some recorders print the demand each half hour on a paper tape whilst others move a pointer round a scale leaving it at the highest point reached.

Where charges are based on kVA it is necessary to take power factor into account. There are a number of ways of doing this amongst which is to record both kWh and $kVA_r h$ from which the kVA and power factor can be computed.

DIVERSITY FACTOR

Within a factory or premises, not all the installed equipment will be working simultaneously.

$$\text{Diversity factor} = \frac{\text{Demand of equipment actually connected at any instant}}{\text{Maximum demand}}$$

LOAD FACTOR

$$\text{Load factor} = \frac{\text{Energy consumed in a given period}}{\text{Energy that would have been consumed had the maximum demand been sustained during that period}}$$

COST OF ELECTRICAL ENERGY

Domestic

Because of the high cost of special metering and the relatively low demand involved the capital charges are recovered from domestic consumers by using a front end loaded tariff with possibly a fixed charge in addition. The use of this type of tariff is best illustrated using an example.

Example 1. Calculate the cost of electricity supplied to domestic premises which have a load factor of 0.05 (5%) and a maximum demand of 10 kW for the following two tariffs:
(a) Fixed charge £5; First 150 kW at 6 p/kWh, all over 150 kWh 2.5 p/kWh. All charges per quarter which is 91 days.

(b) Flat rate of 4 p/kWh

(a) Load factor $= \dfrac{\text{Energy consumed per quarter}}{\text{Maximum demand} \times \text{hours per quarter}}$

$$0.05 = \frac{\text{Energy consumed}}{10 \times 91 \times 24}$$

Energy consumed = 1092 kWh per quarter
Fixed charge = £5
150 kWh at 6 p/kWh = £9
(1092 − 150) kWh at 2.5 p/kWh = £23.55

Total cost = £37.55 Average price per kWh $= \dfrac{3755}{1092}$

$$= 3.44 \, \text{p/kWh}$$

(b) 1092 kWh at 4 p/kWh = £43.68.

Example 2. A family uses 2000 kWh in a winter quarter. The quarterly tariff is:
First 100 kWh cost 8 p/kWh: all over 100 kWh cost 2.4 p/kWh.
The maximum demand is 12 kW.
Calculate: (i) the average cost per kWh
         (ii) the load factor for the quarter.

Industrial

A number of variations on tariffs exist and these are best illustrated using a further example.

Example 3. A factory has a maximum demand of 200 kW, a load factor of 0.4 (40%) and an average operating power factor of 0.7.
Calculate the annual cost of energy and the average price per kWh on the following three tariffs:
(a) Maximum demand charge £20/kW. Running charge 3.5 p/kWh.
(b) Maximum demand charge £17/kVA. Running charge 3.5 p/kWh.
(c) Basic maximum demand charge £20/kW increased by a factor 0.1 (10%) for each 0.1 that the power factor is worse than 0.9. Running charge 3.5 p/kWh.

(a) $0.4 = \dfrac{\text{Energy consumed}}{\text{Maximum demand} \times \text{hours in the year}}$

Energy consumed $= 0.4 \times 200 \times 365 \times 24 = 700\,800$ kWh
Maximum demand charge $= £20 \times 200 = £4000$
Energy charge $= 700\,800 \times 3.5\,\text{p} = £24\,528$
Total cost $= £28\,528$ ($= 2852\,800$ pence)

Figure 3.3

Average cost per kWh = cost/number of kWh used
$$= 2852\,800/700\,800$$
$$= 4.07\,\text{p/kWh}.$$

(b) Power factor = 0.7 = cos φ (see *Figure 3.3*).

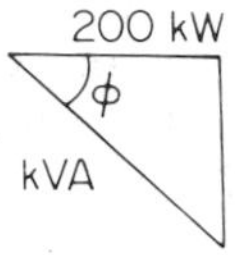

$$\frac{200}{\text{kVA}} = 0.7 \quad \text{Maximum demand in kVA} = \frac{200}{0.7} = 285.71$$

Maximum demand charge = £17 × 285.71 = £4857.07
Energy cost as before = £24 528
Total cost = £29 385.07
Average cost/kWh = 2938 507 p/700 800 = 4.19 p/kWh

(c) The power factor = 0.7. This is 0.2 worse than 0.9 so that the maximum demand charge is increased by a factor 0.2. The maximum demand cost = £20 + (0.2 × £20) = £24/kW
The maximum demand charge = £24 × 200 = £4800
Energy cost = £24 528
Total cost = £29 328
Average cost/kWh = 2932 800/700 800 = 4.18 p/kWh.

Example 4. Factory with an annual consumption of 1 million kWh has a load factor of 0.5. The electricity tariff is £20 per kW of maximum demand plus 3.8 p/kWh. Calculate the average cost per kWh.

Effect on electricity charges of improving load factor. As an example suppose a factory has an energy requirement of 1 million kWh per annum.
The electricity tariff has a maximum demand charge of £15/kW. Let us consider the effect of changing load factors on the maximum demand charge.
(a) With a load factor of 0.25

$$0.25 = \frac{1000\,000}{\text{Maximum demand} \times 365 \times 24}$$

Maximum demand = 456.6 kW
Annual maximum demand charge = £15 × 456.6 = £6849.31
(b) With a load factor 0.75
Using a similar calculation to that above, the maximum demand
= 152.21 kW
Maximum demand charge = £2283.1
Saving = £4566.21
The energy costs will be the same in both cases.
It can be seen therefore that an improvement in the load factor leads to cost saving where the tariff includes a maximum demand charge.
This improvement in load factor may be achieved in a number of ways. Suppose a factory with a metal foundry and a machine shop

both started up simultaneously in the morning. There will be a heavy demand for the first hour or so after which the demand will reduce as furnaces reach working temperature, motors warm up and in winter the workshops reach a comfortable temperature and possibly lighting is switched off. The load factor is low since the heavy demand only lasts for a small proportion of the working day.

By either working a night shift in the foundry or starting it up say two hours earlier than the machine shop, the demand by the foundry can be reduced to a very low value when the machine shop starts running. The maximum demand is thereby reduced and the load factor increased. The total energy used remains at about the same level.

The load factor on commercial premises could be improved by using off peak storage heating instead of direct acting equipment so that the heating load is off when lighting and other services are required for the morning start.

Increasing the number of hours that a particular piece of equipment works improves its load factor. Where several pieces of equipment are used only intermittently ensuring that only one is used at a time improves the load factor.

Effect of power factor improvement.   Where a tariff incorporates a maximum demand charge based on kVA, or has a power factor penalty clause as in Example ‑3(c), savings can be achieved by improving a low power factor. Such a tariff seeks to discourage low power factor since this involves larger currents than necessary to perform a given amount of work. Conductors, transformers and switchgear must therefore be larger than for the same load at unity power factor.

Consider a factory with a maximum demand of 400 kW paying for electricity on a tariff of £12/kVA of maximum demand.

(i) At a power factor of 0.2 lagging. $\phi_1 = 78.46°$ (see *Figure 3.4*)

$$kVA_1 = \frac{400}{0.2} = 2000$$

Maximum demand charge = £12 × 2000 = £24 000

(ii) At a power factor of 0.6 lagging. $\phi_2 = 53.1°$

$$kVA_2 = \frac{400}{0.6} = 666.67$$

Maximum demand charge = £12 × 666.67 = £8000

(iii) At a power factor of 0.9 lagging. $\phi_3 = 25.8°$

$$kVA_3 = \frac{400}{0.9} = 444.4$$

Maximum demand charge = £12 × 444.4 = £5333.3

Thus considerable savings can be made by improving the operating power factor. The improvement may be brought about by the use of capacitors connected across the supply either at the load point itself or in the substation feeding the factory.

A capacitor draws a current which leads on the supply voltage by 90° and therefore has no active component. It is totally reactive and

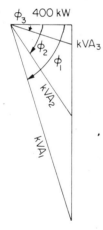

Figure 3.4

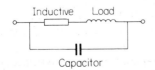

Figure 3.5

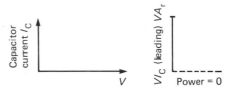

Figure 3.6

multiplying by the supply voltage V gives VI_c, leading volt-amperes reactive (see *Figure 3.6*).

Adding the capacitor to the load does not affect the load power but since the leading and lagging volt-amperes reactive are in direct phase opposition the arithmetic difference may be taken as shown in *Figure 3.7*.

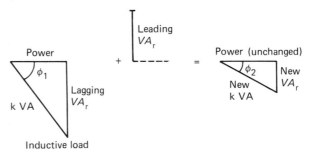

Figure 3.7

New value of VA_r = Lagging VA_r − Leading VA_r
The overall phase angle has been reduced from ϕ_1 to ϕ_2 and the value of kVA has been reduced.

As well as capacitors, there are specially designed motors which can be caused to operate at leading power factors so having the same effect. These are installed to drive loads within the factory.

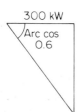

300 kW
Arc cos
0.6

Figure 3.8

Example 5.   An industrial load has a maximum demand of 300 kW at a power factor of 0.6 lagging. Calculate the saving in maximum demand charges and the overall saving if a capacitor is fitted which draws 150 kVA$_r$. The tariff is £12/kVA of maximum demand. The annual capital charges for the capacitor are £600.
[Note that capacitors, like all plant, have to pay capital charges as already discussed. For a capacitor the running charges are considered to be zero since there is virtually no power loss]

$\cos \phi = 0.6$ therefore $\phi = 53.13°$ (using tables or calculator)

$$kVA = \frac{300}{0.6} = 500$$

$$\frac{kVA_r}{300} = \tan \phi = 1.33. \ kVA_r = 1.33 \times 300 = 400$$
(or use Pythagoras' theorem)

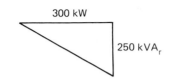

Figure 3.9

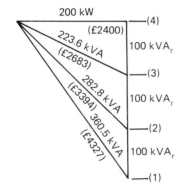

Figure 3.10

Maximum demand charge = £12 × 500 = £6000.
Since the capacitor draws 150 kVA$_r$ leading, the total number of lagging kVA$_r$ will be reduced by this amount to 400 − 150 = 250 kVA$_r$. Redraw the triangle as in *Figure 3.9*.
By Pythagoras' theorem: kVA = $\sqrt{(300^2 + 250^2)}$ = 390.5.
New maximum demand charge = £12 × 390.5 = £4686.
Saving on maximum demand charge = £6000 − £4686 = £1314.
But the annual cost of the capacitor = £600. Therefore the net saving = £1314 − £600 = £714.

Example 6.   Calculate the saving in maximum demand charge if a factory with maximum demand 500 kW at a power factor of 0.7 lagging improves this to 0.9 lagging. The maximum demand charge is £14/kVA.
What is the overall annual saving if the power factor correction equipment costs £800 per annum?

It is usually uneconomic to correct the power factor to unity since the costs of so doing become greater than the savings as the power factor approaches this value.
Consider the case of a factory with a maximum demand of 200 kW at a power factor such that it requires 300 kVA$_r$ lagging. This is condition (1) in *Figure 3.10*.
We will consider the effects of adding capacitor banks to the system in three stages each drawing 100 kVA$_r$ leading. The capital charges on each 100 kVA$_r$ bank are £400 per annum. The maximum demand charge is £12/kVA.
With no capacitors in circuit the maximum demand is 360.5 kVA and the maximum demand charge is £12 × 360.5 = £4327.
Adding one bank of capacitors improves the power factor so reducing the kVA demand to 282.8 at a cost of £3394 (point (2) in *Figure 3.10*).
Adding the next bank of capacitors reduces the kVA demand to 223.6 at a cost of £2683 (point (3) in *Figure 3.10*).
Adding the final capacitor bank brings the power factor to unity when the kVA demand is equal to the number of kW and the maximum demand charge is £12 × 200 = £2400 (point (4) in *Figure 3.10*).
The savings from (1) to point (2) = £933 at a cost of £400 spent in capacitors.

net saving = £533.

From point (1) to point (3) the savings = £1644 at a cost of £800.

net saving = £844.

Correcting to unity, from point (1) to point (4) the saving in maximum demand charge is £1927 at a cost of £1200. Net saving £727.

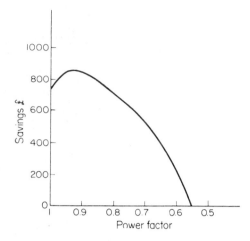

Figure 3.11

Alternatively consider changing from point (3) to point (4). The saving in maximum demand charge = £283 which costs £400 for capacitors.

Figure 3.11 shows a graph of savings against power factors for the above case. The maximum savings are achieved by correcting to a power factor of 0.942. Going beyond this value costs more for capacitors than can be saved on maximum demand charges. Different capacitor and maximum demand charges will alter the power factor for optimum savings.

PROBLEMS FOR SECTION 3

7 A private house contains the following equipment:
   Electric lamps, total power 1100 W
   Television set 200 W
   Electric iron 750 W
   Two electric fires 4000 W
   Washing machine with heater 4500 W
   Food mixer 500 W
   Electric cooker 10 000 W
   Electric kettle 3000 W
   (a) Calculate the diversity factor when (i) 200 W of lighting, the television set and both electric fires are operating, (ii) the cooker is operating at half maximum power, the electric iron and kettle are switched on.
   (b) Calculate the load factor for the installation if in one quarter year 1250 kWh were consumed and the maximum demand was that of 50% of the total equipment listed above.
   (c) Calculate the total electricity charge for the quarter and the average cost per kWh for a tariff: Fixed charge £2.50, First 150 kWh at 2.8 p/kWh, all in excess of 150 kWh, 2.1 p/kWh.

8 The basic charge on a certain tariff is 4.5 p/kWh. There is a fuel adjustment clause whereby the unit charge may be changed by 0.0004 p/kWh in the appropriate direction for every penny the fuel cost changes from £25/tonne. Calculate the cost per kWh when the fuel charges (a) fall to £20/tonne and (b) rise to £26/tonne.

9 During eight successive half hours the energy meter connected to the supply lines to a commercial premises was noted to make the following advances:
100, 150, 125, 175, 200, 150, 125 and 130 kWh respectively. Assuming that these are the highest noted during the year, determine the maximum demand charge on a tariff of £20/kW of maximum demand.

10 Why is there no cost advantage for domestic consumers to practice (a) power factor improvement or (b) load factor improvement?

11 Calculate the values of the quantities missing from the following table.

	kW	kVA	kVA$_r$	power factor
(i)	150	?	75	?
(ii)	150	?	?	0.65
(iii)	?	?	300	0.7

12 A load of 500 kW at a power factor of 0.64 lagging is to be corrected by the addition of capacitors in three stages. Each stage of correction adds 200 kVA$_r$ leading to the system. Calculate (i) the kVA loading on the system and (ii) the overall power factor for (a) the uncompensated load (b) with one stage of compensation (c) with two stages of compensation and (d) with three stages of compensation.

13 A factory has a maximum demand of 650 kW at a power factor of 0.65 lagging. The maximum demand charge is £18/kVA. What would be the annual saving on the electricity bill if a capacitor was fitted which drew 600 kVA$_r$ leading?

14 A factory uses 850 000 kWh per annum. The load factor = 0.3.
The tariff is £20/kVA of maximum demand + 4 p/kWh.
Calculate (a) the maximum demand in kW and kVA assuming a power factor of 0.7 lagging and (b) the new M.D. in kVA and the size of capacitor required (in kVA$_r$) if the electricity bill is to be reduced by £2000 per annum.

15 A factory consumes 1.2 million kWh per annum. It has a load factor of 45% and operates at an average power factor of 0.65 lagging.
  (a) Calculate the total electricity charges on a tariff of £14/kVA of maximum demand plus 1.6 p/kWh.
  (b) Calculate the savings to be made by improving the power factor to 0.9 lagging. The capital charges on power factor correcting equipment are £5/kVA$_r$.

(c) Calculate the additional saving in maximum demand charge by improving the power factor to unity. What additional cost is incurred in the provision of correcting equipment?

Would this final improvement be economically sound?

16 A factory with a foundry and a machine shop has a maximum demand of 750 kW and a load factor of 0.2. The electricity tariff is £20/kWh of maximum demand + 4 p/kWh of energy consumed.

(a) Calculate the annual cost of electricity on this tariff.

(b) To save money, a two-shift system of working is introduced. The foundry starts work earlier in the morning whilst the machine shop commences rather later in the day. By this means the maximum demand charge is reduced, since all the equipment will not be switched on simultaneously at the beginning of each day. This working pattern increases the number of kWh consumed by 5% but increases the load factor to 0.35. Calculate the saving in electricity charges.

4 Materials and their applications in the electrical industry

Aims: At the end of this chapter you should be able to:

Compare aluminium and copper as conductor materials.
Explain the effect on cable rating of ambient temperature and apply grouping factors.
Use tables to determine cable sizes for specified loadings and situations.
Describe the construction of printed circuit boards and explain the need to limit the current density in the conductors.
Explain why heat sinks are necessary for many electronic devices.
Explain why dielectrics fail in service.
Describe the constructional details of a range of cables.
Explain why core losses occur in iron-cored coils and transformers.
Compare the properties of permanent magnet steels with those of soft ferro-magnetic materials.
Describe the required properties for core materials for use with a range of frequencies.

CONDUCTOR MATERIALS FOR OVERHEAD LINES AND UNDERGROUND CABLES

The best electrical conductor known is silver but this is far too expensive and rare to provide all the conductor material required by the electrical industry. Next in order of conductivity come copper and aluminium and these are the most important current carrying materials used in cable and line manufacture. Other materials used are cadmium-copper, phosphor-bronze and for some high-voltage low-power links, galvanised steel.

The conductivity of both copper and aluminium falls very rapidly with very small additions of alloying elements so that they are generally used pure. In the case of aluminium the mechanical strength is improved by using a stranded conductor with steel strands at the centre.

Figure 4.1 shows a stranded conductor and the make up will be 1 strand of steel plus 6 strands of aluminium; 7 strands of steel plus 12 strands of aluminium, or by adding a further layer of aluminium strands, 7 of steel plus 30 of aluminium.

The conductivity of the whole is taken as that of the aluminium alone since steel has a very high impedance and the current flows almost exclusively in the aluminium.

The total strength of the cable is normally 50% greater than that of the same conductivity copper cable. The result of using steel cores

1+6+12

Figure 4.1

is to produce cables which are smaller for a given tensile strength than copper or, although larger for a given resistance, much stronger than copper. The table details some of the electrical and mechanical properties of copper and aluminium.

Property	Copper	Aluminium
Weight	87 200 N/m³	26 700 N/m³ (Aluminium 0.306 times copper)
Resistivity	1.73×10^{-8} Ωm (0.975 as good as silver)	2.87×10^{-8} Ωm (Aluminium 1.64 times copper) (0.585 as good as silver)
Strength	Ultimate 320 MN/m²	Ultimate 144 MN/m² (Aluminium 0.45 times copper)
Flexibility	When annealed, quite good. Used hard for cables and overhead lines when it must be stranded to give the required flexibility.	Very flexible, can be used solid in cables.
Jointing	Soldered ferrules for cables. Compression joint on overhead lines.	Compresson type on overhead lines. Crimped lugs for cable terminations. Can be soldered or welded using special fluxes but generally more difficult than copper.
Resistance to corrosion	Excellent. Virtually none in most circumstances	Poor when in contact with other metals and in particular copper and copper bearing alloys. Special sleeves and fittings required. On overhead lines there is limited deterioration, much of the original 1933 grid is still operational in its original form.

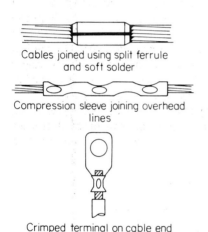

Cables joined using split ferrule and soft solder

Compression sleeve joining overhead lines

Crimped terminal on cable end

Figure 4.2

PERFORMANCE OF INSULATING MATERIALS

Loss angle δ

In order to prevent current leaking away from cables and other conductors, they need to be suspended on or surrounded by insulating materials. For example, overhead power lines are insulated from earth by air along their length and by porcelain or glass insulators from the steel supporting towers, whilst inside, leakage from power cables is prevented by enclosing them entirely in plastic materials or paper. Such insulating materials are known as *dielectrics*.

A dielectric is a material which has a very high resistivity in comparison with a conductor such as copper and aluminium. However no dielectric is perfect and there will always be some current leakage. This leakage current, flowing in its resistance, is in phase with the voltage and so represents a power loss. In addition, consider how a capacitor is formed. Any conducting surface close to another but separated from it by an insulator has capacitance. A capacitor for use in electronic circuits often comprises two very thin aluminium foils separated by waxed paper or mica dielectric. Now since the current-carrying conductor of an overhead line or within a power cable is near to earth and probably near to the return conductor of the circuit, capacitance is created from conductor to conductor and from each conductor to earth, the insulating material around the conductor being the dielectric.

The current which flows in the circuit because of its capacitance has a value given by the circuit voltage divided by the capacitive reactance of the system.

$$I_C = \frac{V}{X_C} = \frac{V}{1/\omega C} = V\omega C \text{ amperes}$$

The current flowing in the circuit because of the imperfect dielectric $= I_R$ amperes.

The working equivalent circuit and the quantities are shown in *Figure 4.3(a)*. The phasor diagram may be constructed as in *Figure 4.3(b)*. I_C leads the circuit voltage by 90° whilst I_R is in phase with the voltage. The total circuit current I_T leads the voltage by an angle ϕ. Since I_R is usually very small, ψ approaches 90°. The angle δ is the difference between 90° and ϕ, and is known as the loss angle since it is only present because of the dielectric loss. In a perfect capacitor the current would lead the voltage by exactly 90°.

Good cable dielectrics have very small values of loss angle, being only a few minutes of arc. Since insulation resistance decreases rapidly as its temperature increases, it follows that the loss angle δ

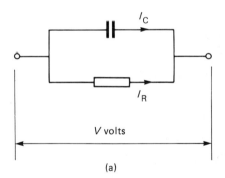

(a)

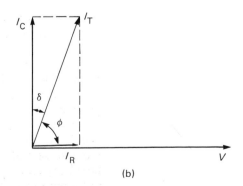

(b)

Figure 4.3

increases with temperature (I_R increases). As δ increases, the loss increases, giving rise to more heat and further temperature rises. It can be envisaged that this could lead to failure if the temperature were allowed to rise too far. This is particularly important in very high voltage cables where the heat produced by this effect becomes very nearly as great as that due to conduction losses in the conductor (I^2R losses). See also page 50, 'Heating of cables'.

Breakdown voltage If a sufficiently high voltage is applied to any insulating material it will break down and become conducting. Particular materials have uses as dielectrics for a particular range of voltages. The breakdown strength of a dielectric depends on its thickness but is not proportional to it. This means that if a sample of insulating material is subjected to an increasing voltage it will eventually puncture and possibly set on fire. Let us suppose that the sample tested was 1 mm in thickness and failure occurred at 5000 V. Using twice the thickness would not result in a breakdown voltage of 2×5000 V but more likely only 7000 or 8000 V. Increasing the thickness further gives even less increase, 3 mm thickness breaking down at perhaps only 10 000 V.

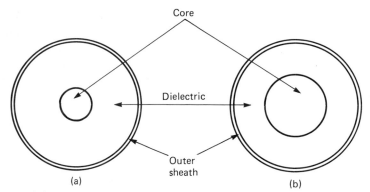

Figure 4.4

In cable design it is found that the diameter of the conductor has a great influence on breakdown voltage. The smaller this is, the more likely the cable is to fail. In *Figure 4.4*, cable (a) will fail at a substantially lower voltage than cable (b) despite the fact that the radial thickness of the insulation is greater than in (b). See also *Figure 1.4* and observe that the conductor in the extra-high voltage cable has been made larger than is required to carry the current by the inclusion of an oil duct through its centre. This has at least as much to do with its high voltage performance as does the provision of oil to improve the insulating properties of the paper dielectric.

INSULATING MATERIALS FOR CABLES For underground cables operating at up to 3.3 kV, the most common insulating material is polyvinyl chloride (PVC). Street mains have either copper or aluminium conductors. These are PVC-insulated, then three or four such conductors are laid up to form a cable which is sheathed overall in PVC, and surrounded with steel wire armouring if required. See *Figure 1.4*.

(a)   (b)

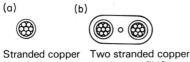

Stranded copper   Two stranded copper
conductor   conductors PVC
                     insulated
PVC insulation   Central single strand
                     earth wire

Figure 4.5

House wiring usually comprises copper conductors which are insulated with PVC (see *Figure 4.5*).

Polychloroprene (PCP) is used to insulate copper conductors in farming installations where resistance to ammonia and other such corrosives is required.

The insulating material in mineral-insulated cables is magnesium carbonate which spaces solid copper conductors within a copper tube. It has a number of advantages amongst which is the ability to operate at red heat – a property which no other cable-insulating material can match.

The insulation for high voltage power cables is predominantly paper with oil or gas filling under pressure. At extra-high voltage it is essential that the dielectric be absolutely uniform with no small air bubbles or particles of dirt present. It would be at these slight imperfections, since they would be weaker than the perfect dielectric, that breakdown would occur. A small arc would form, burning the insulation producing carbon and other conducting combustion products. The process would then be continuous, the arc getting progressively larger until the cable failed explosively. In paper-insulated cables the conductors are wrapped in many layers of paper tape and then oil impregnated under vacuum to completely soak the paper and fill all the minute gaps present with oil. Using this construction, all air bubbles and dirt are excluded. When operating, many oil-filled cables are kept under pressure so that any tendency for the oil to migrate out as the cable heats up is prevented. Within the last 10 years plastic insulation called cross-linked polyethylene (XLPE) has been developed for use at voltages up to 275 kV. The construction is very similar to the solid paper type, polyethylene replacing the paper, but there is no oil and therefore no provision for oil ducts within the cable. It promises to be cheaper to install and maintain than the solid paper type but as yet there is very little of it actually in service.

CABLE CONSTRUCTION

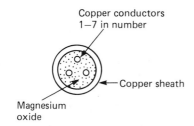

Copper conductors
1–7 in number

Magnesium
oxide

Copper sheath

Figure 4.6

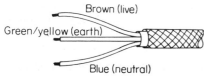

Brown (live)

Green/yellow (earth)

Blue (neutral)

Figure 4.7

At low voltages the single-stranded conductor with PVC insulation is most commonly used. These are drawn into steel or plastic conduit (*Figure 4.5(a)*).

In domestic and commercial situations the twin copper conductor with an earth wire, PVC insulated, is often used (*Figure 4.5(b)*). These cables are either run on the surface or buried in the walls. In hazardous situations, mostly in industry, the mineral insulated cable (MICS) is used and this is also run on the surface or buried (*Figure 4.6*).

For connecting appliances to the supply at low voltages via a plug and socket, the three-core flexible cable is employed (two cores only if the appliance can run earth-free, as in the case of double-insulated equipment). This comprises fine stranded copper wire which can withstand flexing, PVC insulation and belt, covered overall with a braided protective fibre or fabric sheath (*Figure 4.7*).

For frequencies higher than that of the mains, the flat twin or coaxial construction is adopted. (See 'Conductor screening and coaxial cable', below.)

The construction of extra-high voltage cables is illustrated in *Figures 1.4* and *4.8*.

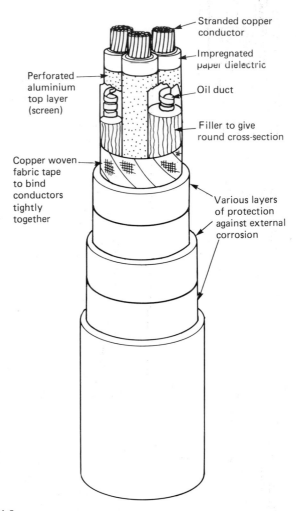

Stranded copper conductor

Impregnated paper dielectric

Perforated aluminium top layer (screen)

Oil duct

Filler to give round cross-section

Copper woven fabric tape to bind conductors tightly together

Various layers of protection against external corrosion

Figure 4.8

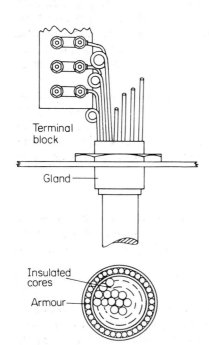

Terminal block

Gland

Insulated cores

Armour

Figure 4.9 PVC insulated multi-core cable

Multi-core cables are used where many circuits are required to run over the same route between a pair of terminal points, for example for the control of equipment, metering and indication and transmission of condition information such as temperature, pressure, position and power consumption. Power to the equipment concerned is usually fed through a separate cable to minimise the risk of damage to low voltage equipment should a fault occur.

The cables comprise the required number of cores with fibre fillers if necessary to make up a circular cross section, with a belt of insulation and armouring if required. Each core is numbered or otherwise identified along its length.

The heavy current types in common use are:

1 Stranded copper conductors with paper or fabric insulation with overall paper wrapping, lead covering, steel wire armoured with an anti-corrosion sheath. The cable is terminated at each end in a compound filled box.

2 Stranded copper conductors with PVC insulation, PVC belt, armoured with PVC anti-corrosion sheath overall.

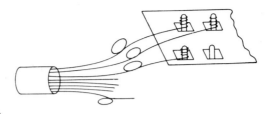

Figure 4.10

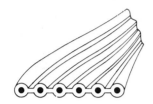

Figure 4.11

Each conductor is terminated in a soldered or crimpled lug which is bolted to a terminal block from which connections to equipment are made.

Lighter types comprise the required number of single strands of fine copper wire insulated with PVC and covered with a belt of PVC. End connections may be crimped as before or bound and soldered to terminal posts.

Multi-core cables for telecommunications have pairs of fine insulated wires twisted together along their length to form circuits, two pairs then being twisted together. Many such groups of four conductors make up a complete cable.

For the interconnection of printed circuit boards, flexible flat cables may be used. The spacing between the conductors is the same as that for the contacts in the edge connectors. Cables with 50 or even more conductors are available.

MATERIALS USED IN ELECTRONIC CIRCUITS

Because of its good conductivity and mechanical strength copper is used extensively to connect electronic components either in the form of insulated stranded wires or strip bonded to an insulating board in the form of a printed circuit. The insulation used is of polythene, polystyrene, polyurethane enamel or air where conductors can be suitably spaced.

Permanent connections between components and interconnecting copper are usually soldered. Where components or boards of components must be capable of removal from an equipment, possibly for repair, some form of sliding contact system is required. The plug and socket for connection of domestic electrical equipment is an example of such a system.

CONDUCTOR SCREENING AND COAXIAL CABLE

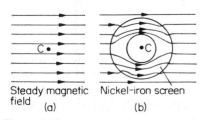

Steady magnetic field   Nickel-iron screen
(a)   (b)

Figure 4.12

A conductor carrying current sets up a magnetic field around it, the strength of which is proportional to that current. This magnetic field may affect other circuits in its vicinity and to prevent this circuit screening is necessary.

Where a component is to be screened from a steady magnetic field it is surrounded with a material with low magnetic reluctance. In *Figure 4.12* the component to be screened is marked C. Without the screen the magnetic field will affect the component. With the screen in position the magnetic lines of force take the low reluctance path which is through the nickel-iron, so leaving the component in a position of zero flux.

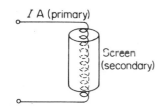

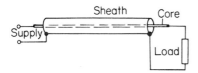

Figure 4.13

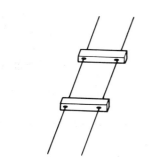

Figure 4.14

Figure 4.15

Figure 4.16

Where alternating current is involved the conductor carrying the current, and hence producing the alternating magnetic flux, is surrounded by the screen. The arrangement is in effect a transformer, the conductor being the primary and the screen a short circuited secondary.

The alternating flux induces a voltage in the screen and a current flows which sets up an opposing magnetic flux. The two fluxes are very nearly equal so that the magnetic field outside the screen is virtually zero.

The magnetic field around a cable feeding a circuit is eliminated by using the coaxial construction. This is particularly important at very high frequencies when non-screened cables can cause considerable interference with other circuits.

Current is supplied to its load along the core of the cable and is returned to the supply through the sheath which completely surrounds the core. Since the go and return currents are in opposite directions, the magnetic fields produced are in opposite directions and completely cancel each other outside the sheath.

The cable may be formed using a solid core with a copper tube for the sheath with solid insulation (MICS) or by using a solid or stranded core with a braided outer sheath for flexibility when the insulation is generally polythene.

For many applications, air is the best insulation. However, air alone is not always practicable, since the conductors have to be kept to their correct spacing. At frequencies up to 200 MHz open wires may be used, employing small spacing elements of conducting material as shown in *Figure 4.15*. An alternative construction is shown in *Figure 4.16*, in which two parallel conductors are lightly insulated and maintained in their positions using a thin plastic web. This construction is used for audio frequencies, feeding loudspeakers from an amplifier, for example, and at radio frequencies, often being used to connect a VHF (FM) aerial to a receiver. The spacing between the conductors is critical, this having a pronounced effect on the performance of the equipment.

At frequencies up to 3000 MHz the coaxial construction may be used. A compromise between using solid insulation and air is achieved by using either plastic foam containing a substantial amount of air in the form of bubbles or by using a polythene spiral wound on the inner core, when again the insulation is substantially air. This is shown in *Figure 4.17*.

Figure 4.17

The coaxial construction is also used at power frequencies, as illustrated in *Figure 1.4*.

PRINTED CIRCUITS

The electrical connections between circuit components must have low resistance and this is often achieved by using insulated copper wire or bare copper or aluminium strip. The working components are then generally individually soldered to the conductors. In equipments with many components there are several problems to overcome. One is that of the dry joint, when the soldering looks sound but in fact has not made good contact and the joint has a high resistance. Another problem with individual hand wiring is that it is difficult to reproduce a circuit accurately in successive equipments. There may be errors in connections and slightly different lengths and routing for wiring taken. Circuit response is thereby affected. Fault finding can be very difficult and it may be necessary to unsolder connections to do tests.

Mounting components on an insulating board to a fixed pattern is one solution to these problems. This is especially true now that valves have been largely replaced by transistors and integrated circuits in which the heat generated is relatively small. The fixed pattern ensures repeatable values of circuit resistance and capacitance.

The board must provide mechanical support for the components and the means for their interconnection. The layout of components is much clearer than with individually wired equipment and the values of the components are often marked on the board. Servicing of the equipment is simpler since faulty boards may be replaced to get the equipment working and the components on the faulty boards replaced at a central workshop.

Modern rigid boards are made of phenolic or epoxy resins reinforced with woven glass fabric or paper. They are available up to about one metre square. Flexible boards are available and these are made from polyester film or for high temperature work from PTFE or polypropylene possibly reinforced with glass fibre.

Printed circuit boards employ copper foil conductors securely cemented to one of the above laminates. Holes are drilled through the foil and board and the component tails are fixed by soldering. *Figure 4.18* shows a very simple board with two components.

To construct a printed circuit board it is necessary to draw a master circuit diagram. Larger areas of foil have to be allowed for parts of the circuit which will carry the heaviest currents. The master circuit can be drawn directly on to the board in the case of a prototype or it can be printed on using a photographic process when large numbers of identical circuits are required.

Two processes are available for the production of the final circuit.

1 The subtractive method. A board is obtained which has foil completely covering one side. The diagram is drawn or printed on the foil using acid-resisting paint and the board is dipped into acid which dissolves the copper which is not protected. Only the required circuit is left on the board and the necessary holes can be drilled for circuit assembly.

2 The additive method. Insulating board is used with no foil covering. The circuit is drawn or printed on the board using conducting paint. Copper is deposited from a plating bath over the treated areas. Other additive methods involve the use of metallic powder and heat or foil strips and mechanical force.

The current carrying capacity of the foil is determined largely by the permissible temperature rise. Charts showing current carrying

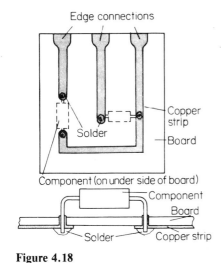

Figure 4.18

capacities and cross-sectional areas for different increases in temperature are consulted when circuits are being designed. Foils vary in thickness from 0.035 mm to 0.106 mm and in width from 0.25 mm up to several millimetres.

As an example a strip 0.07 mm thick and 0.76 mm wide can carry 3.5 A allowing a 40°C rise in temperature but only 2 A allowing a 10°C rise, both from 15°C.

The cross-sectional area of this strip = 0.07 × 0.76 = 0.053 mm²

Allowing a 40°C rise, the permissible current density $= \dfrac{3.5}{0.053}$ = 65.8 A/mm²

Allowing a 10°C rise, the permissible current density $= \dfrac{2}{0.053}$ = 37.6 A/mm²

Increasing the strip width gives a less than proportional increase in current carrying capacity so that current for other strips cannot be deduced by proportion.

The current densities obtained in these very small conductors are considerably greater than those which can be achieved with mains cables since the insulation is very much thinner and the surface area available for cooling is greater per given volume.

To connect the board into the main circuit either plug and socket or edge connectors are used.

Figure 4.19 shows the form of the edge connector. Where large boards are used there may be over 100 connections to make so that considerable force may be necessary to insert a multi-pin plug into its socket or the board into its edge connector. When using such force it is difficult to determine whether a correct match of the contacts has been made or whether in fact some of them are being badly deformed or crumpled up.

There are several patent ways of achieving low insertion pressure followed up by clamping of the board in position. The situation is eased considerably by the choice of contact materials. For example the pressure needed between gold or platinum contacts to obtain very low contact resistance is only 0.03 times that necessary to obtain the same resistance between brass contacts. The value for nickel and silver is 0.3 times that for brass.

Gold is generally used for low voltage connections of this type since in addition to the low contact pressure required, it does not oxidise or corrode. Even a very thin oxide film can cause a virtual open circuit unless the voltage employed is large enough to break it down. The gold, generally less than 50×10^{-6} m thick, is plated over silver or nickel, the latter combination being especially good in hostile environments.

Interconnection between boards is achieved using insulated wires which are grouped into multi-core cables or merely bunched and neatly carried round the chassis of the equipment. Great care has to be taken to ensure that the capacitance between cores in such close proximity does not affect the circuit performance, that all the connections are sound and that any one circuit does not affect any other as already discussed in the section 'Conductor screening and coaxial cable'.

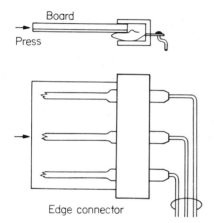

Figure 4.19

CIRCUIT DAMAGE DUE TO OVERCURRENTS

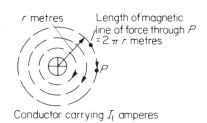

Figure 4.20

Consider a single conductor carrying a current of I_1 amperes and the resulting magnetic field as shown in *Figure 4.20*.

The magnetising force at point P is given by: $H = \dfrac{I_1}{l_1} = \dfrac{I_1}{2\pi r}$ A

where l_1 is the length of the magnetic line of force through P. Since the magnetic flux density $B = \mu_0 H$ tesla (strictly in vacuum, but considered true for air and solid insulators),

$$B = 4\pi \times 10^{-7} \times \frac{I_1}{2\pi r} = 2 \times 10^{-7} \times \frac{I_1}{r} \text{ tesla}$$

A second conductor carrying I_2 amperes running parallel to the first and through point P will suffer a force given by $F = BI_2l_2$ newtons. The arrangement is shown in *Figure 4.21*.

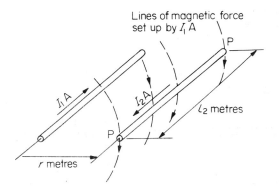

Figure 4.21

So that for $l_2 = 1$ m

$$F = 2 \times 10^{-7} \times \frac{I_1 I_2}{r} \text{ newton/metre length.}$$

Action and reaction being equal and opposite this must also be the value of the force on the first conductor. With currents in opposite directions as shown the forces are of repulsion. In single-phase a.c. and direct current circuits $I_1 = I_2$ and under normal conditions the forces involved are small. However, when a short circuit fault occurs the currents and forces can be extremely large, sufficient in fact to disrupt the cable or conductor system.

Consider two conductors 4 mm apart carrying (a) a normal current of 20 A and (b) a fault current of 2500 A.

(a) $F = 2 \times 10^{-7} \times \dfrac{20 \times 20}{4 \times 10^{-3}} = 0.02$ newtons per metre run ($l_2 = 1$)

(b) $F = 2 \times 10^{-7} \times \dfrac{2500 \times 2500}{4 \times 10^{-3}} = 312.5$ N/m run

In three-phase circuits each conductor is situated in a magnetic field produced by the currents in the other two conductors and the same effect is produced.

In addition to the forces involved the heating effect of large currents must be considered.

Power loss $= I^2R$ watts.

Increasing the circuit current from 20 A to 2500 A will increase the power loss by a factor $(2500/20)^2$ or 15 625 times. It is easy to envisage the effect this will have on the circuit insulation. Within a few cycles of an a.c. supply, plastic insulation will melt.

HEATING OF CABLES

Since all conductors in normal service have resistance, the passage of current along the conductor gives rise to a voltage drop. The voltage drop is in phase with the current producing it so that the power loss is equal to the product of voltage drop and current flowing.

Consider one core: Volt drop $= IR$ where $R =$ resistance of one core.

Power loss = volt drop $\times$ current $= IR \times I = I^2R$ watts per core.

This power heats the core producing a rise in temperature of the conductor and hence of its insulation. The heat is lost to the surrounding air or ground. The temperature rises until the rate of dissipation from the insulation is equal to the rate of heat production in the conductor. Unfortunately good electrical insulators are often good heat insulators and it is quite possible to produce heat in the conductor at such a rate that the temperature at which equilibrium would be reached would also damage the insulation. In addition, the leakage current through the insulation must be considered. This produces heat as does the current flowing in the conductor. For example a cable operating at 240 V which has an insulation resistance of 240 megohms will allow $240/240 \times 10^6$ amperes to flow and the power loss will be 0.24 mW.

Figure 4.22 shows that as the temperature of a cable insulating material rises its insulation resistance falls. Hence at a higher temperature the leakage current is greater so that more heat is produced in the insulation. An unstable situation can be created where more heat increases the temperature which lowers the resistance of the insulation so allowing more leakage current to flow which produces more heat etc. The cable eventually fails by burning under these circumstances. See also page 40, 'Loss angle'.

PVC becomes soft at temperatures in excess of 80°C and the conductors tend to migrate through the insulation eventually touching each other on earth. An upper working temperature limit of 70°C for PVC is set for this reason.

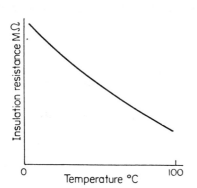

Figure 4.22

PERMITTED VOLT-DROPS IN CABLES

The 16th edition of the IEE Regulations, regulation 525–01–02, states that the size of conductor feeding a load shall be such that the volt drop at the load does not exceed 4% of the nominal voltage, not taking into account supply fluctuations or current surges due to motor starting and the like.

The following data is taken from Tables 4D1A and 4D1B of the 16th edition of the IEE Regulations.

Current-carrying capacities and associated volt drops for unsheathed single-core PVC insulated cables in conduit on a wall

Conductor cross-sectional area	Single-phase circuit (Two Cables)	
(mm²)	Current carrying capacity (A)	Volt drop per ampere per metre (mV)
1.0	13.5	44
1.5	17.5	29
2.5	24	18
4	32	11
6	41	7.3
10	57	4.4
16	76	2.7

Correction factors for ambient temperatures when using fuses to BS 88 or BS 1361 (HRC fuses) (Regulation 523–01 and Appendix)

Ambient temperaure (°C)	25	35	40	45	50	55
Factor	1.06	0.94	0.87	0.79	0.71	0.61

EFFECT OF AMBIENT TEMPERATURE ON RATING

The amount of heat which can be conducted away from a cable depends on the temperature difference between the cable and its surrounding medium. A conductor which is at 60°C in an environment which is at 60°C cannot dissipate any heat. The same conductor in an environment at 5°C will dissipate a large amount of heat.

The upper working temperature of the insulation is fixed at a value at which it will not deteriorate rapidly and at which its mechanical properties will be unimpaired. In the previous section we saw that this was 70°C for PVC. The heat which can be dissipated therefore depends on the temperature of the surrounding medium, or *ambient temperature.* For a particular current, the heat to be dissipated can be decreased by decreasing the resistance of the conductors. This will involve increasing their cross-sectional areas.

A cable with a particular cross-section which can carry 20 amperes with the ambient temperature at 25°C might need to be replaced with one which has twice the original cross-section to carry this current in an ambient temperature of 60°C.

The increase in area depends not only on the ambient temperature but on the type of circuit protection employed. Where this is a high- rupturing-capacity fuse, the increase is less than when a re-wireable semi-enclosed fuse is used. The difference is due to the degree of overloading which the different devices allow before operating. (We will consider HRC fuse protection only, see table of values in the previous section.)

Example 1. A 240 V, single-phase circuit has a full load current of 15 A and is protected by a 15 A, HRC fuse to BS 88. The circuit conforms to Tables 4D1A and B. The circuit runs in a region with ambient temperature 45°C. Determine the required cross-sectional area of the cable conductors. Calculate the volt drop in the circuit if the run is 20 m in length. Is this satisfactory? [Reg 525-01-02]

From the table the correction factor for ambient temperature = 0.79

The cable has to be rated at $\dfrac{15}{0.79} = 19\,A$

(Note that although the cable must be capable of carrying 19 A, it will carry only the full load current of 15 A. We must use this value in our volt-drop calculations.)

From Table 4D1A/B we see that a 1.5 mm^2 cable will carry 17.5 A, which is not quite enough. It will therefore be necessary to use the next size whiich is 2.5 mm^2 in cross-section.

The allowable volt drop is 4% of 240 V which is 9.6 V.

The volt drop for 2.5 mm^2 cable is 18 mV/A per m. With 15 A this is $15 \times 18\,mV = 0.27\,V/m$.

With a length of 20 m the total volt drop = $20 \times 0.27 = 5.4$ V.

This is less than 9.6 V and so is regarded as satisfactory.

CABLES IN DUCTS OR IN CLOSE PROXIMITY

Cables which are in ducts or are laid close to each other suffer mutual heating: one cable, becoming warm, heats its neighbours. Where cables touch each other, the number of free paths by which heat can escape to the environment is reduced. The general mass of cables may raise the temperature of the surrounding air or ground. For these reasons it is necessary to reduce the heat generated by the cables and this is done by increasing the cross-sectional area, as already discussed under 'Effect of ambient temperature on rating'. A factor is applied which may be determined by experiment or obtained from sources such as the IEE Regulations or Electrical Research Association data.

A small part of Table 4B1 from the 16th edition of the IEE Regulations is quoted here.

Correction factors for groups of more than one circuit of single-core cables, or more than one multicore cable 'enclosed' or 'clipped direct' conditions.

	Correction factor							
Arrangement of cables	*Number of circuits or multicore cables*							
	2	3	4	6	8	10	12	14
Enclosed in conduit or trunking, or bunched and clipped direct	0.8	0.7	0.65	0.57	0.52	0.48	0.45	0.43

Example 2. Estimate the required cross-sectional area of the conductors to Table 4D1A/B when they are protected by HRC fuses with a rating of 10 A for the following circuits:
(a) a single circuit situated in an ambient temperature of 25°C
(b) six circuits run together inside trunking, the ambient temperature being 25°C and,

(c) six circuits run together inside trunking, the ambient temperature being 50°C.

(a) The ambient temperature correction for 25°C = 1.03

The cable must be capable of carrying $\dfrac{10}{1.03}$ A = 9.71 A

Using Table 4D1A/B we see that a 1 mm^2 conductor has a current-carrying capacity of 13.5 A so this will be adequate.
(b) The grouping factor for 6 circuits = 0.57. Again the ambient temperature correction = 1.03.

The cable must be capable of carrying $\dfrac{10}{1.03 \times 0.57}$ A
= 24.7 A.

This will require the use of 4 mm^2 conductor which can carry up to 32 A.

Example 3.   A single-phase circuit is protected by an HRC fuse to BS 88 with a rating of 20 A. The nominal circuit voltage is 230 V. The load is 30 m from the supply point and fuse. Using Table 4D1A/B determine the cross-sectional area of a suitable conductor given that there are no temperature or grouping factors to consider.

A 2.5 mm^2 conductor is rated at 24 A and would seem appropriate on first consideration. Investigate the volt drop along its length.
We are allowed 4% of 230 V = 9.2 V.
The 2.5 mm^2 circuit has a volt drop of 18 mV/A/m.
Volt drop = 20 A × 18 mV × 30 m = 10.8 V. This is too great and the conductor size must be increased.
Try 4 mm^2 conductor. Volt drop = 11 mV/A/m.
Volt drop in this circuit = 20 A × 11 mV × 30 m = 6.6 V.
This is less than 9.2 V and is therefore satisfactory. We have in fact used cable rated at 32 A to carry 20 A in order to satisfy the volt drop requirements.

Example 4.   Select suitable conductors from Table 4D1A/B for the following circuits. The currents quoted refer to the rating of the protective HRC fuse.
(a) 15 A, circuit length 10 m, nominal circuit voltage 250 V. No temperature considerations required
(b) 15 A, circuit length 10 m, nominal circuit voltage 100 V. No temperature considerations required.
(c) 30 A, circuit length 25 m, nominal supply voltage 200 V. Ambient temperature 35°C.
(d) 10 A, circuit length 5 m, nominal supply voltage 240 V. Ambient temperature 50°C.
(e) With the details as in (d) above but with 4 identical circuits contained in a conduit.

HEAT SINKS

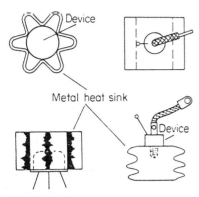

Figure 4.23

The characteristics of semi-conducting devices such as rectifiers and transistors change considerably with a rise in temperature so much so that the circuit in which they are connected may cease to operate in the manner intended. For example, a rectifier will become conducting in both directions if it is made hot enough. Often a device is permanently damaged and the original characteristics are not re-established by cooling. Where excessive currents have been drawn due to these changes, other components in the circuit may have been damaged.

Insulation deteriorates at high temperature and the printed circuite board itself or connecting wires can suffer damage. With plug-in components the contacts may suffer since contact springs can become soft and ineffective so introducing a high resistance and more heat into the circuit.

Heat sinks are metal clip or screw-on additions to a device which effectively increase the surface area and so aid cooling.

Two types of heat sink are shown in *Figure 4.23*. These may be fan cooled in extreme cases.

THE INSULATION OF OVERHEAD LINES

The material most commonly used for overhead line insulators is porcelain. The insulators are shaped from the raw material: a mixture of clay, finely ground feldspar and silica in water, are dried, dipped in liquid glaze and then fired at very high temperature. The glaze forms a glass-like coating providing a surface to which dirt cannot readily stick. It also improves the strength of the insulator so that fracture is more difficult.

One disadvantage of porcelain is that when the glaze is chipped by a power arc or by missiles projected by vandals, water can soak into the body of the insulator. This causes it to become conducting and an electrical discharge takes place which trips out the line while the heat produced dries out the insulator. Such a fault is immensely difficult to find, the line tripping out each time there is rain.

Glass is an alternative material. Toughened glass insulators have a higher breakdown strength under electrical stress than porcelain and if damaged shatter completely rather like the windscreen of a motor vehicle when it is hit by a stone. Missing insulators leave gaps which are easily spotted during one of the regular inspections which are made. The construction is such that although the insulator shatters the line cannot fall down.

The glass is toughened by heating it to the softening point and then cooling it fairly rapidly. This causes the surface to become hard whilst the centre is still plastic. As the centre cools and sets it tries to contract so pulling in on the outer layers. Before the insulator can fracture, these internal stresses have to be overcome and the force required can be six times that required to break ordinary glass.

Suspension insulators are made of porcelain or glass whilst pin insulators are almost invariably made of porcelain (see *Figure 1.3*).

Insulators must withstand mechanical and electrical stresses. Heavy lines must be held off the ground whilst the electrical potential of the line is considerably above that of earth. Even in wet weather the insulator must function and for this reason it has sheds or skirts to keep at least part of the surface dry in almost any weather conditions.

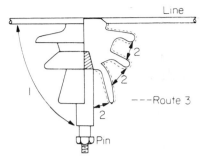

1 Dry flashover distance
2 Wet flashover distance
   (exposed surfaces wet & conducting)
3 Dry leakage distance

Figure 4.24

Figure 4.24 shows a pin type insulator supported at the bottom with the line at the top. The steel pin is at earth potential. An electrical breakdown or flashover can occur in one of three ways:

1 A dry flashover can occur. This means that an arc will form round the insulator from line to pin along the route 1 in *Figure 4.24*.

2 A wet flashover can occur. When the top surfaces of the insulator are wetted by rain they become conducting so that the sheds are so disposed to keep the undersides dry. A flashover can occur along the route 2 in *Figure 4.24*.

3 Current can leak from the line over the surface of the insulator along the route 3 in *Figure 4.24*.

Atmospheric pollution causes dirt to accumulate on the surface of the insulator making it partly conducting and a flashover more likely. Fog wets the insulator overall causing surface leakage to occur especially when smoke and sulphur oxides are also present. Sea spray carried by the wind also has the same effect.

EDDY CURRENT AND HYSTERESIS LOSSES

When a conducting material is situated in a region of changing magnetic flux a voltage is induced in that material and a current flows. By Lenz's law the direction of this current is such as to create an opposing magnetic flux. *Figure 4.25* shows a single turn wound on a conducting core. The current in the winding is increasing. The core flux is therefore increasing and voltages are induced in (a) the current-carrying coil itself, this voltage restricting the rate of rise of the current, (b) the core material, this voltage driving what are called eddy currents which heat the core since it has resistance, (c) any other coil or conducting material in close proximity to the core, this being the basis of the transformer.

The induced voltage in case (a) is a self-induced voltage, the coil possessing self-inductance whilst in cases (b) and (c) we have mutually induced voltages, there being mutual inductance between the coil and core and any other conductor in the vicinity.

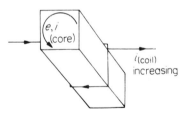

Figure 4.25

From Faraday we know that the induced voltage

$$e = \frac{\text{Change in flux linkages}}{\text{Time taken for the change}}$$

Using symbols: $e = \dfrac{\Delta \Phi}{\Delta t}$ the Greek Δ (delta) being read as 'the change in'.

Considering the core as a single turn, when an alternating current flows in the coil in *Figure 4.25*, the time interval for a current to change from zero to a maximum or from a maximum to zero etc., with the corresponding changes in core flux, is a function of frequency. The greater the frequency the smaller is the time interval for the change.

Hence, the time interval $\Delta t \propto \dfrac{1}{f}$

So that $e \propto \dfrac{\Delta \Phi}{\dfrac{1}{f}}$ $e \propto \Delta \Phi f$ volts.

Now since $i = \dfrac{e}{R}$

and power $= ei$ watts, power $= e \times \dfrac{e}{R} = \dfrac{e^2}{R}$ watts.

Where R = resistance offered to the current flowing by the core material.

The eddy current power loss $\propto \dfrac{(\Delta\Phi f)^2}{R}$

In addition, when ferrous cores are used there is a hysteresis loss which is due to the reversal of the magnetic field in the core material. Energy is required to orientate very small groups of atoms called Domains so as to create a magnetic field in one direction during a positive-going half-cycle and then to re-orientate them with opposite polarity during the negative-going half-cycle.

The power loss is a function of the maximum working flux density; how many of the domains need to be orientated to give the required field strength; and of the number of times this is carried out per second; the frequency.

Hysteresis power loss $\propto \Delta B \times f$ watts.

MAGNETIC MATERIALS

Certain materials such as iron, cobalt and nickel can be made very strongly magnetic, either permanently when they are known as permanent magnets, or temporarily when surounded by a current-carrying coil, when they are called electro-magnets.

In *Figure 4.26* a coil of N turns is shown carrying a current of I amperes. The product of $N \times I$ is known as the magneto-motive force and it is this m.m.f. which is creating the magnetic field which links with the coil. The m.m.f. divided by the length of the coil l is called the magnetising force which has symbol H.

$$H = \frac{NI}{l} \text{ amperes per metre}$$

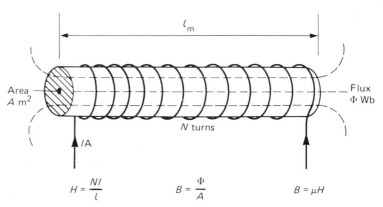

Figure 4.26

This magnetising force sets up a magnetic field of value Φ webers. The density of the field is expressed as webers per square metre, symbol B.

$$B = \frac{\Phi}{A} \text{ webers per square metre or tesla (T)}$$

The relationship between B and H is called permeability, symbol μ, such that:

$$B = \mu H$$

A material which has high permeability will have a strong magnetic field created in it in response to a small amount of excitation, that is to say, for a small product of $N \times I$. *Figure 4.27* shows a set of curves of flux density plotted against magnetising force for different materials. Observe that the magnetic field strength increases rapidly at first but eventually levels off at values between 1.1 and 1.5 tesla. Where the curves level off, it is said that the materials are saturated. Generally we are interested in the region of the curves before saturation occurs, since here there is maximum magnetic field effect for the minimum energy input to the exciting coil. This means that in this region the permeability is large. If we move into saturation, the magnetising force H goes on increasing for no increase in flux density. This gives a low value of permeability.

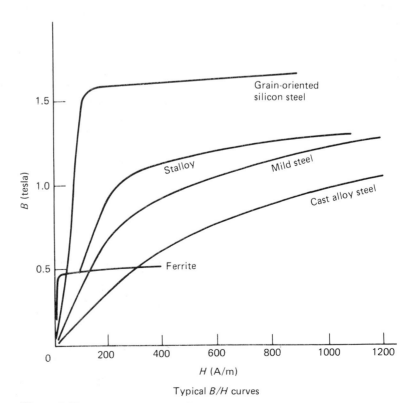

Typical *B/H* curves

Figure 4.27

For example, in a particular iron sample it takes a value of $H = 200$ A/m to set up a flux density $B = 1$ tesla.

Since $B = \mu H$, $1 = \mu \times 200$, $\mu = 1/200 = 5 \times 10^{-3}$

Moving into saturation, it requires a value of $H = 1500$ A/m to set up a flux density $B = 1.3$ tesla.

$$1.3 = \mu \times 1500, \quad \mu = 1.3/1500 = 0.866 \times 10^{-3}$$

Let us now examine the properties of some materials used in devices in which magnetic fields are required.

1 *Ferrite.*   The original 'lodestone' was used by mariners who first discovered that this material would always point in a particular direction. A lump of this material, the earliest form of compass, enabled them to navigate. Lodestone is magnetite or ferrous ferrite. These are mixtures of iron oxides. (Rust is an iron oxide as is the black scale which forms when iron is heated.) Ferrites are manufactured at the present time from iron oxides together with oxides of manganese, nickel and zinc. Ferrites have high permeabilities but can only work up to about one half of the maximum flux density of steel. One of their advantages over steel in certain circumstances is that the body of the material has a very high resistance to electrical currents flowing in them, so that eddy current losses will be very small. Some of the ferrites can be permanently magnetised, when they become extremely difficult to de-magnetise.

2. *Soft iron.*   This is very nearly pure iron. It is very soft and can be readily bent or formed into any required shape. It may contain a little silicon (up to about 4%) to increase its resistance to internal current flow. If silicon iron is rolled out from a fairly thick sheet to a very much thinner sheet whilst cold and then heated up and slowly cooled to anneal it, it becomes what is known as 'grain-oriented'. When it is used as an electromagnet, provided that the magnetic field runs in the same direction that the sheet was rolled, it exhibits greater permeability than ordinary soft iron. This, together with its high resistance makes it ideal for use as transformer core laminations.

3 *Soft alloys.*   Alloys of nickel and pure iron with possibly a small addition of molybdenum and copper are very soft and are known under their trade names, amongst which are Stalloy and Mumetal. They have a permeability which lies between that for nearly pure iron and the grain-oriented variety. Since they contain no silicon, their resistances are low. Their high permeability and low resistance make them suitable as screening materials. Mumetal was so named after the Greek letter μ, which is used as the symbol for permeability.

4 *Hard alloys.*   These alloys are so hard as to be unrollable when cold, and cannot be filed or bent. They are cast into blocks from the molten material and can only be shaped using a high-speed grinding wheel. Such materials would be found in tools such as files, some spanners and screwdrivers etc. Particularly when used as a magnetic material, they will be made from various mixtures of iron, nickel, cobalt chromium, copper and tungsten. Permeability is fairly high but is not a prime requirement. What is required is the ability to retain their permanent magnetism in any adverse conditions.

5 *Magnetic dust.* Particles of nickel, iron molybdenum or pure iron are individually covered with an insulating material, mixed with an insulating resin and then pressed into any required shape. They may then be operated as electro magnets without any possibility of current flowing through the material from particle to particle. These materials have lower permeabilities than the irons since there is less active material present (a lot of space being taken up by the filler), but they are almost immune from eddy current losses.

6 *Air.* Although not generally considered to be a magnetic material, there are a number of applications where an electro magnet with an air core is the only possibility. It has very low permeability ($4\pi \times 10^{-7}$), but suffers no internal losses.

MINIMISATION OF LOSSES

In electrical machines, working at frequencies up to a few kilohertz, high values of magnetic flux density are required so that a small cross-sectional area of magnetic material may be used to give the required value of flux.

Flux density × area of magnetic core = total flux

Using a small area minimises the weight and hence the cost of the equipment.

The eddy current losses in the core are reduced by increasing its resistance. This is achieved by the use of laminations. A lamination is a very thin sheet of material, less than 0.4 mm thick, which is carefully cleaned and varnished or anodised on one side. Many laminations are pressed together to form the required cross-section of the core. An alternative, but less effective method uses thin sheets of paper between the laminations. *Figure 4.28* shows a simple arrangement. Eddy currents can now no longer flow as shown in *Figure 4.25* but are constrained to very small loops within each of the insulated laminations. The use of 4% silicon steel increases the resistance of the individual laminations whilst the magnetic strength may be increased for a given coil and current by the use of grain-oriented steel.

Laminations of soft nickel–iron alloy at thicknesses of less than

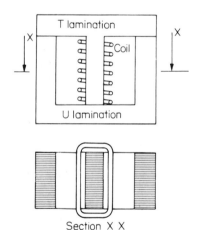

Figure 4.28

0.1 mm may be used at frequencies of up to about 10 MHz. For frequencies higher than this, since the eddy current losses are proportional to the frequency squared, these may be considerable even with the thinnest laminations. Thus dust or ferrite cores are used up to about 150 MHz in communications equipment. In ultra-high frequency applications, air cores may have to be resorted to, since air suffers no hysteresis or eddy current losses. *Figure 4.29* shows a small coil with a ferrite or dust core which is threaded. It may be screwed into or out of the coil to adjust the inductance value.

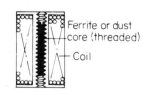

Figure 4.29

PERMANENT MAGNETS

Permanent magnets are made from the hard alloys. They are heated to a high temperature, shaped and allowed to cool inside a coil carrying a large direct current. They then become permanently magnetised. They are used, for example, in permanent magnet moving-coil d.c. instruments, loudspeakers and small electric motors. They may become demagnetised if they are heated to a high temperature either by an external heat source or if they are operated in the presence of a high-power alternating magnetic field from another piece of equipment, when the alternating field will cause eddy currents to flow within the permanent magnet (as already described under 'Eddy current and hysteresis losses'). This will produce internal heating and possible demagnetisation. One solution to this problem is the use of permanent magnet ferrites. These are not so strong as the metal types but do not suffer eddy current losses so do not heat up in the presence of external fields.

MAGNETIC SCREENING

If a coil carrying alternating current is enclosed within a cylinder (a screen) of soft alloy (Mumetal), it becomes what is in effect a transformer with the screen as its secondary, the coil being the primary. By Lenz's law, the current flowing in the screen will produce a magnetic field in opposition to that of the coil. The two magnetic fields cancel and there is no magnetic effect outside the screen. See 'Conductor screening and coaxial cable' and *Figure 4.13*. Since Mumetal has a high permeability, the magnetic field is strong and hence the eddy currents are large even though the cross-sectional area of the screening material is small. As the frequency increases, the e.m.f. induced in the screen increases and the eddy currents will increase in magnitude. The screen therefore becomes more and more effective as the frequency increases. Above 20 kHz the Mumetal screen may be replaced by one of copper or aluminium since although their permeabilities are less, the increasing frequency produces the necessary increase in eddy currents for good screening.

PROBLEMS FOR SECTION 4

5 Although silver is the best electrical conductor, it is not used in cable manufacture. Why is this?

6 How is the low tensile strength of aluminium compensated for in the construction of overhead lines?

7 Hard drawn copper is not very flexible. How is flexibility built into cables and lines with copper cores?

8 Why would PVC insulation not be used in very hot situations? Suggest a suitable alternative type of insulation.

9 Calculate the volt drops in the following circuits using Table 4D1A/B.
   (i) c.s.a. 2.5 mm^2 Current 20 A   length of run 3 m
   (ii) c.s.a. 1.5 mm^2 Current 12 A   length of run 5 m
   (iii) c.s.a. 4 mm^2 Current 25 A   length of run 10 m

10 Determine suitable conductor sizes for the following circuits (Table 4D1A/B). The nominal supply voltage is 240 V.

Fuse rating (BS 88)	Temperature of run (°C)	Number of circuits in a single conduit	Length of run (m)
(a) 10	35	1	25
(b) 10	40	3	6

| (c) 15 | 50 | 2 | 40 |
| (d) 30 | 45 | 6 | 8 |

11 A single-phase circuit carrying 50 A has a power loss of 1 W per metre run due to conductor heating. There is a force between the conductors of 0.1 N per metre run due to the magnetic effect. Calculate the power loss and the force between the conductors when a short circuit occurs and the current rises to 5000 A.

12 How is the choice of cable affected by the type of fuse employed to protect the circuit?

13 A cable has an insulation resistance of 20 MΩ and a capacitance of 0.01 μF. The supply voltage is 66 000 V at a frequency of 50 Hz. With reference to *Figure 4.3*, calculate the value of I_R and I_C and hence determine the size of the loss angle δ. What is the power loss in the cable?

14 What is the purpose of oil filling under pressure in a paper-insulated power cable?

15 What are the possible advantages to using MICS cable as opposed to PVC insulated cable in an industrial situation?

16 Why is it necessary to use coaxial cables when interconnecting equipments operating at very high frequencies?

17 Describe the types of cable available for use at radio frequencies.

18 What are the advantages of using printed circuit boards in a television receiver, as compared with an individually-wired chassis?

19 Describe two processes which are available for the manufacture of printed circuit boards.

20 A printed circuit board has one circuit formed from foil 0.035 mm thick and 0.5 mm wide. The permitted current density for a 40°C temperature rise is 50 A/mm^2. Calculate the maximum permissible current in the circuit.

21 Calculate the current densities in the following printed circuit board conductors:
   (a) 0.07 mm $\times$ 1 mm carrying 5 A.
   (b) 0.106 mm $\times$ 2 mm carrying 4 A.

22 What is the function of a heat sink?

23 Name the two principle insulating materials used in overhead line construction.

24 What is the difference between a pin insulator and a suspension insulator?

25 Why is flashover on an overhead line insulator more likely to occur on a rainy day than on a dry day? What features are built into the insulator to minimise the occurrence of flashover?

26 Suggest suitable core materials for the following, giving reasons:
   (i) a power transformer with a rating of 500 MW in service 24 hours every day,
   (ii) a portable power transformer rated at 500 W for use only a few hours in a week (*Hint:* (i) and (ii), think about what importance the losses have in each of the situations),

(iii) a coil operating at 100 MHz,

(iv) a coil operating at 1000 MHz.

27 In what situation might a permanent magnet ferrite be used in preference to a metal alloy permanent magnet?

28 A coil is to be screened to prevent interference with adjacent circuits. Suggest suitable materials for the screening can for frequencies (a) 15 kHz, (b) 10 MHz.

29 The hysteresis loss in a transformer core is 75 W when it is operating at 50 Hz. What would be the expected hysteresis loss if the frequency were doubled to 100 Hz, provided that the core flux remained constant?

30 The eddy current loss in the transformer in Problem 29 is also 75 W at 50 Hz. If the core flux is held constant, what would be the expected value of eddy current loss at 100 Hz?

5 Transformers

Aims: At the end of this chapter you should be able to:

Sketch the types of cores and windings found in transformers.
State an area of use for each of the following types of transformer:
(i) single-phase double-wound, (ii) three-phase, (iii) auto, (iv) current, (v) potential.
Explain how instrument transformers are used.

Faraday discovered that whenever a change in magnetic flux is associated with a coil of wire a voltage is induced in that coil. The value of the induced e.m.f. is proportional to the number of turns and to the rate of change of magnetic flux in webers per second.

$$e = N \frac{d\Phi}{dt} \text{ volts}$$

Alternating voltages of any desired value may be obtained by using the transformer which employs this principle. Voltages need to be changed between the points of generation and the consumer several times in order to arrive at the most economical levels for transmission and distribution. Generation is carried out at voltages between 11 kV and 25 kV whilst major transmission voltages are 275 kV and 400 kV. Domestic consumers are supplied at about 240 V.

PRINCIPLE OF ACTION OF THE TRANSFORMER

Figure 5.1 shows the general arrangement of a transformer with the secondary open circuited. There are two coils, generally known as the primary and secondary, wound on an iron core. The iron core is made up of laminations which are about 0.3 mm thick. These have been rolled to the correct thickness, acid cleaned, polished and varnished or anodised on one side, and then made up into the correct core form.

When an alternating voltage V_p is applied to the primary coil a small magnetising current flows which sets up a magnetic flux in the iron core. This alternating flux links with both the primary and secondary coils and with the iron of the core inducing voltages in each. The voltage E_p induced in the primary coil opposes the applied voltage V_p according to Lenz's law. The difference between V_p and E_p is very small. The voltage induced in the iron core causes eddy currents to flow so giving rise to the production of heat. Dividing the core into well insulated laminations increases the resistance so minimising these currents and the associated loss. The eddy current and hysteresis losses due to alternating magnetisation must be provided by the power source.

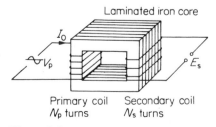

Laminated iron core

Primary coil
N_p turns

Secondary coil
N_s turns

Figure 5.1

Finally the voltage E_s induced in the secondary winding is used to supply the load.

EMF EQUATION

Let the maximum value of core flux be Φ_m webers and the frequency f hertz.

The time taken for the flux to change from $+ \Phi_m$ to $- \Phi_m$ is $\dfrac{\tau}{2}$ or $\dfrac{1}{2f}$ seconds.

Since $e = N\dfrac{d\Phi}{dt}$ volts

the average e.m.f. induced in the primary winding

$$= N_p \times 2 \Phi_m \div \frac{1}{2f} = 4N_p\Phi_m f \text{ volts.}$$

Where N_p = number of turns on the primary winding.

For a sine wave, the r.m.s. value is 1.11 times the average value. Therefore $E_p = 4.44 N_p \Phi_m f$ volts and if all the magnetic flux set up by the primary winding links with the secondary, $E_s = 4.44 N_s \Phi_m f$ volts, where N_s = number of turns on the secondary.

Therefore $\dfrac{E_s}{E_p} = \dfrac{N_s}{N_p}$

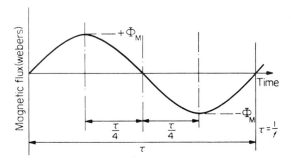

Figure 5.2

THE SINGLE-PHASE TRANSFORMER

The single-phase transformer has two windings, a primary and a secondary. These are usually wound on a laminated iron core. The windings may be arranged as shown in *Figure 5.1* but, since it is important to cause as much as possible of the magnetic flux which is set up by the primary to pass through, or link with, the secondary, the windings are more commonly either arranged concentrically or interleaved, as shown in *Figure 5.3*. The interleaved winding is sometimes referred to as a *pancake construction*. One type of core uses lamination stampings as shown in *Figure 5.4*. The two 'E' stampings plus the two end-closing stampings form an element of the core. Many more of these elements are added to make up the correct core thickness.

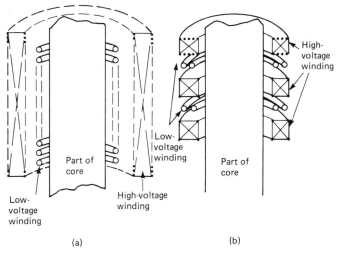

Figure 5.3

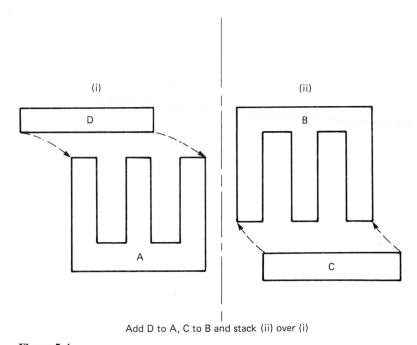

Add D to A, C to B and stack (ii) over (i)

Figure 5.4

A complete air-cooled single-phase transformer is shown in *Figure 5.5*. Even with the coils so close together, there is always some leakage flux, that is to say magnetic flux created by one of the coils which does not in fact link with the other. This occurs partly because, where the edges of laminations meet, the grain-structure of the iron is discontinuous; the shapes of grains and the direction in which they have been rolled are different in one lamination from those in the lamination it touches, unless immense care is taken when cutting the lamination from the large roll of strip material. This gives rise to an increase in reluctance at the joints, causing

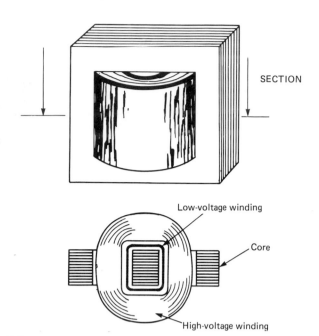

Low-voltage winding

Core

High-voltage winding

Figure 5.5

leakage flux and an increase in iron losses. Another reason for leakage flux is that, however the windings are arranged, there is a slight gap between primary and secondary and between the windings and the core. Again, some leakage flux may be created. Leakage flux can cause interference with other circuits, inducing voltages which cause hum in communications networks.

Where low core losses combined with virtually zero leakage flux are sought, a core formed from a continuous strip of grain-oriented steel may be used. The strip is wound on a mandrel of the correct size and subsequently heat-treated to remove the stresses set up by this process. The windings are added either in the form of one of the arrangements already described or as shown in *Figure 5.6*, which shows the toroidal transformer as used in high quality audio equipment. The path of the flux follows the line of the strip, there is no tendency to leak away and full use is made of the properties of the grain-oriented material. There are no joints across which flux must pass. Another use of this construction is in the three-phase transformer shown in *Figure 5.7(a)*.

Some typical uses for single-phase transformers are:

1   Power supplies for electronic equipment. Taking a supply at 240 V, the transformer will step this down to a lower value where it will be rectified and smoothed to give a direct supply to, for example, amplifiers, instrumentation, and oscillators.

2   To provide a supply in workshops at reduced voltage for illumination and power tools. The reduced voltage is used for safety reasons and may be at 110 V or 25 V. The 25 V supply is often derived using a centre-tapped secondary. This point is earthed when the output wires are at ±12.5 V with respect to earth.

3   As an isolating transformer, when neither of the output wires is connected to earth. This means that a user touching either of the

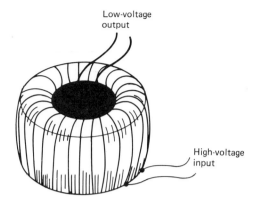

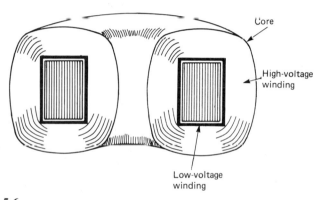

Figure 5.6

secondary wires would not receive an electric shock since there is nowhere for currents which would flow through the body to return to.

4   Electric arc welding. The primary is often connected across two lines of a three-phase supply and so is at 415 V. The secondary provides an output at between 90 V and 110 V when no welding is taking place. This falls to about 25 V once the arc is struck. The use of two lines for the input spreads the load over two phases of the supply. Where several welders are used they should not all be connected to the same two lines.

THE THREE-PHASE TRANSFORMER

Three-phase transformers are essentially three single-phase transformers wound on a single core. The windings are of concentric or pancake type. The windings are arranged on a laminated iron core with either three limbs or five limbs. These are shown in *Figure 5.7*. Notice that in *Figure 5.7(a)* continuous strip is used to form the yoke. When it is completely formed it is cut in half. This is called a 'cut core'. After carefully cleaning the cut and adding the windings, the two halves are re-united. Very little reluctance is sacrificed since

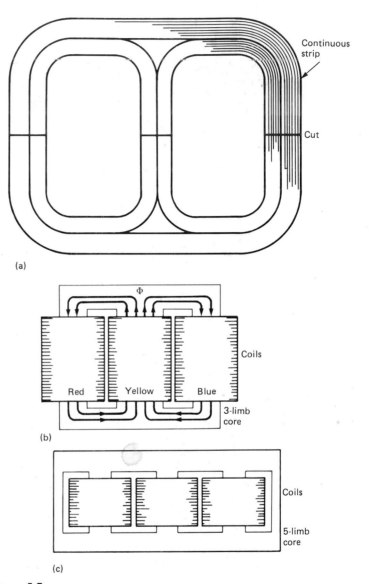

Figure 5.7

at the join the grain all runs in the original direction. Each lamination is being virtually restored to its uncut state.

With the three-phase system, provided that the currents are balanced over the phases, we have seen that the sum of the three currents is zero at any instant. (See *Figures 1.8*, *1.9*, *1.11* and *1.12*). Now, since each of the three currents produces a magnetic field, it follows that at any instant the sum of the fluxes should also be zero. In *Figure 5.8*, consider the instant at which the yellow current and hence the yellow flux is at its maximum positive value, while each of the other two currents has one half of its maximum value, but negative. The fluxes in the red and blue limbs will have a magnitude one half that in the yellow limb and be in the opposite direction. In *Figure 5.7(b)* the maximum flux is shown upwards in the yellow limb, it splits into two halves, one of these is shown downwards in

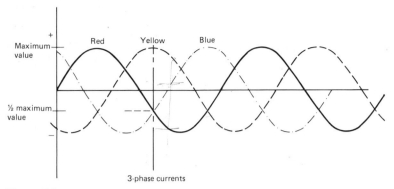

Figure 5.8

the red limb and the other half is shown downwards in the blue limb. At any instant in time this balance of fluxes will obtain.

If the phase currents are not equal, one of the limb fluxes will be of greater magnitude than the sum of the other two. The out-of-balance flux will pass out of the three-limb core and into the surrounding medium. This can be prevented by providing extra limbs. In the five-limb core there are paths for out-of-balance flux. When the load is balanced these have no effect; when it is out of balance, no flux needs to leave the core.

Three-phase transformers are used at all stages of transmission and distribution to create the necessary voltages (see Section 1, 'Transmission and distribution of electrical energy'). Because of the necessity to reduce the transport weight of the very largest transformers, those associated with generators − stepping up from 25.6 kV to 400 kV − are often of the auto type (see under 'The auto-transformer' below). Three-limb and five-limb transformers are employed. All others are double-wound, usually with three limbs.

The primary may be either star-connected or delta-connected, but the secondary is almost invariably star-connected to provide an earthing point. *Figure 5.9* shows a schematic arrangement of a delta-star three-phase transformer.

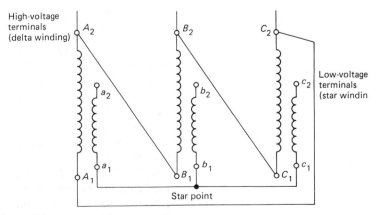

Figure 5.9

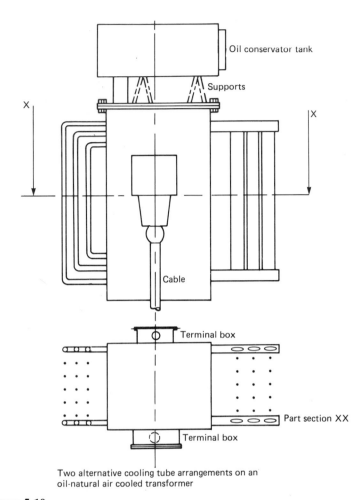

Two alternative cooling tube arrangements on an
oil-natural air cooled transformer

Figure 5.10

TRANSFORMER COOLING

Transformers suffer eddy current and hysteresis losses in their cores
and copper losses (I^2R) in their windings. The heat generated must
be removed or eventually the transformer will become hot enough to
fail. For the smaller ratings, natural air cooling is sufficient. The
transformer is surrounded by air and, due to the heat produced,
convection currents of air are created which carry the heat away.
Provided that free passage of air can be maintained, all will be well.

For higher ratings and in particular for higher voltages, the transformer is immersed in oil within a steel tank. The oil fills all the
small gaps in the insulation, creating a uniform dielectric (see
Section 4, 'Insulating materials for cables'). In addition, the oil will
carry away heat either by natural convection or by forced circulation by an external pump. The oil is cooled in one of three ways:
(i) by natural air convection over external oil-filled tubes (*Figure
5.10*), (ii) by forced air circulation using fans (*Figure 5.11*) or (iii) as
in (ii) except that a water–oil cooler is used, the water being drawn
through many small tubes over which the oil is pumped. The water
pressure is maintained lower than that of the oil so that any leakage
is of oil into the water rather than water into the oil.

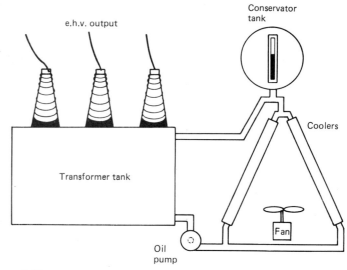

Figure 5.11

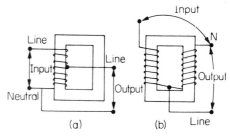

Figure 5.12

(a)   (b)

THE AUTO-TRANSFORMER

The auto-transformer has only one winding. Part of this winding is common to both primary and secondary which are therefore both electrically and magnetically linked.

Figure 5.12 shows the possible arrangements of coils and core. The inputs and outputs are reversible providing for voltage increase or decrease. Considering the transformers to be ideal, i.e. ignoring all losses, the simplified circuits shown in *Figure 5.13* may be drawn.

Figure 5.12(a) shows the current directions in the primary and secondary of a double wound transformer. When the secondary is in fact part of the primary, the current in the secondary section becomes $(I_s - I_p)$ as shown in *Figure 5.13(b)* and the cross-sectional area of this section may be reduced so saving copper.

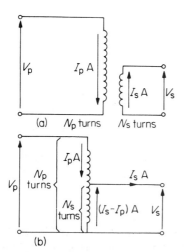

Figure 5.13

Example 1. A single phase auto-transformer has a ratio 500 V:400 V and supplies a load of 30 kVA at 400 V. Calculate the value of current in each section of the winding. Assume ideal operation.

30 kVA at 500 V requires $\dfrac{30\,000}{500} = 60$ A

30 kVA at 400 V requires $\dfrac{30\,000}{400} = 75$ A

Current in the secondary section of winding $= 75 - 60 = 15$ A.

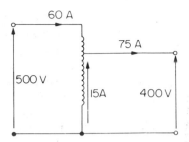

Figure 5.14

ADVANTAGES OF THE AUTO-TRANSFORMER OVER THE DOUBLE-WOUND TRANSFORMER

1 *Less copper is required.* The volume of copper in a winding is proportional to the number of turns and to the cross-sectional area of the wire used, which is in turn proportional to the current to be carried.

Therefore volume of copper $\propto NI$.

For a double-wound transformer

Volume of copper $\propto N_pI_p + I_sN_s$ and assuming ideal operation $N_pI_p = I_sN_s$ Therefore Volume of copper $\propto 2N_pI_p$

For an auto-transformer (see *Figure 5.14*)

$$\text{Volume of copper} \propto N_s(I_s - I_p) + (N_p - N_s)I_p$$
$$\propto N_sI_s + N_pI_p - I_pN_s - I_pN_s$$
$$\propto N_sI_s + N_pI_p - 2I_pN_s \text{ and since } N_sI_s = N_pI_p$$
$$\propto 2N_pI_p - 2I_pN_s$$

$$\frac{\text{Volume of copper in the auto-transformer}}{\text{Volume of copper in the double-wound transformer}} = \frac{2N_pI_p - 2I_pN_s}{2N_pI_p}$$

$$= 1 - \frac{N_s}{N_p}$$

Transposing
Volume of copper in the auto-transformer

$$= \left(1 - \frac{N_s}{N_p}\right) \times \text{volume of copper in a double-wound transformer.}$$

When N_s approaches N_p (V_s approaches V_p) the saving in copper is greatest.

Example 2. Compare the volume of copper in a single-phase auto-transformer with that in a double-wound transformer for ratios (a) 400 V:300 V (b) 400 V:50 V.

(a) Volume of copper in the auto-transformer $= \left(1 - \dfrac{300}{400}\right) \times$ volume in the double-wound transformer $= 0.25$ times

(b) Volume of copper in the auto-transformer $= \left(1 - \dfrac{50}{400}\right) =$ 0.875 times that in the double-wound transformer.

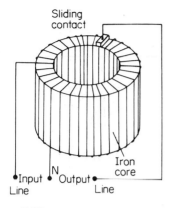

Sliding contact

N

Input Line • • Output Line

Iron core

Figure 5.15

2 *The weight and volume of the auto-transformer is less.*

3 *The auto-transformer has a higher efficiency* and suffers less voltage variation with changing load due to the better magnetic linkage between the primary and secondary sections of the winding.

Auto-transformers are used to interconnect the 400 kV, 275 kV and 132 kV sections of the British grid system. When transporting very large high voltage transformers by road the weight has to be carefully considered and quite often, before major power system construction is commenced, roads and bridges have to be specially reinforced to carry such loads.

It is also worth noting that an increase in efficiency of only 0.5% means a reduction in losses of 2500 kW in a 500 000 kW transformer.

Auto-transformers are also used to reduce the voltage supplied to induction motors and to increase the voltage supplied to discharge lamps during their starting periods.

4 *A continuously variable output voltage is obtainable.* Using the arrangement shown in *Figure 5.15* a continuously variable output voltage may be obtained.

DISADVANTAGES OF THE AUTO-TRANSFORMER

1 *Since the neutral connection is common to both primary and secondary*, earthing the primary automatically earths the secondary. Double-wound transformers are sometimes used to isolate equipment from earth.

2 *If the secondary suffers a short circuit fault*, the current which flows will be very much larger than in the double-wound transformer due to the better magnetic linkage. There is more risk of damage to the transformer and circuit due to heating and the mechanical forces set up between current-carrying conductors.

3 *A break in the secondary section of the winding* stops the transformer action and the full primary voltage will be applied to the secondary circuit.

CURRENT TRANSFORMERS

The current required to give full-scale deflection of a d.c. moving-coil ammeter is very small being typically only a few milliamperes. When large currents are to be measured a shunt or bypass resistor is used in conjunction with the meter. Alternatively a moving-iron meter may be used.

In either case the meter coil is at the potential of the circuit in which the current is being measured. With very large currents the size of the conductor and meter terminals will be large and the internal wiring of control and metering panels made very unwieldy. An alternative using an external shunt is shown in *Figure 5.16*. This involves individual calibration of each such meter with its shunt to allow for the resistance of the connecting wires.

Where alternating currents are involved a shunt cannot be used since the proportion of the current which flows in the meter will depend on its impedance, which varies with frequency. A small change in frequency would upset the calibration of the meter.

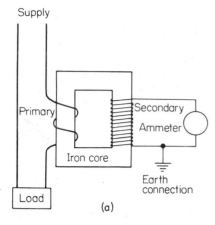

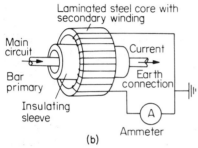

Figure 5.17

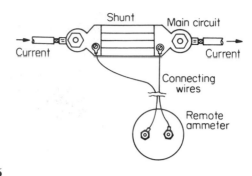

Figure 5.16

These problems are overcome by the use of current transformers which isolate the meter from the main circuit and allow the use of a standard range of meters giving full-scale deflections with 1, 2 or 5 A irrespective of the value of current in the main circuit. Two types of current transformer are shown in *Figure 5.17*.

Figure 5.17(a) has a wound primary while *Figure 17(b)* has a bar primary.

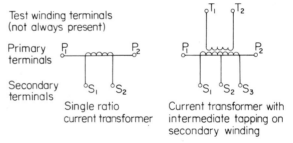

Figure 5.18

Figure 5.18 shows typical terminal markings for current transformers.

The primary of the current transformer is connected in series with the load on the circuit, replacing an ammeter, and has an extremely small voltage drop. The core flux is therefore small. The value of the primary current is determined entirely by the load in the main circuit and not by the load on its own secondary which is typically between 2.5 VA and 30 VA.

As in the power transformer $I_s'N_p = I_sN_s$

Transposing $I_s = I_s'\dfrac{N_p}{N_s}$

Since the core flux is small, the balancing current I_s' may be considered to be equal to I_p.

$I_s = I_p\dfrac{N_p}{N_s}$ (Closely)

The magnetic fluxes set up by the m.m.f.s. I_pN_p and I_sN_s may individually be quite large but are very nearly equal and are in opposite directions.

Example 3. A current transformer has a primary winding of 2 turns and a secondary winding of 100 turns. The secondary winding is connected to an ammeter with a resistance of 0.2 Ω. The resistance of the secondary of the current transformer is 0.3 Ω. The value of current in the primary winding is 250 A. Calculate: (a) The value of current in the CT secondary

(b) The potential difference across the ammeter terminals

(c) The total load in VA on the CT secondary.

(a) $I_s = I_p \dfrac{N_p}{N_s} = 250 \times \dfrac{2}{100} = 5\,\text{A}.$

(b) Potential difference across the ammeter terminals
$= I_s \times$ resistance of the meter
$= 5 \times 0.2$
$= 1\,\text{V}.$

(c) Total resistance of the secondary circuit $= 0.2 + 0.3 = 0.5\,\Omega.$
Since 5 A is flowing, the total induced e.m.f. $- 5 \times 0.5$
$= 2.5\,\text{V}.$

Total VA $= 2.5 \times 5$
$= 12.5\,\text{VA}.$

Example 4. A current transformer has a bar primary (1 turn). The secondary is connected to an ammeter which indicates 4.3 A when the current in the main circuit is 344 A. Determine the number of turns on the CT secondary.

The secondary of a CT must never be open circuited whilst the primary is carrying current since under these conditions $I_s N_s$ will be zero. In a power transformer this would cause I_s' to fall to zero so that the core flux would remain at its normal value. In the CT the primary current is determined by the load in the main circuit and therefore does not fall when the secondary CT circuit is disconnected.

The flux set up by the primary m.m.f. will be unopposed and will link with the secondary inducing a large voltage in it. This may be a danger to life and to the insulation within the CT. The large flux will also cause an increase in the hysteresis and eddy current losses in the core with subsequent heating and further damage to the insulation. The CT may well be ruined.

The secondary of the CT is earthed to prevent its potential rising above that of earth due to the capacitance between the secondary and the high voltage primary. Also in the event of an insulation failure between primary and secondary, the earth connection would allow fault current to flow which should cause the circuit to be isolated. (Electricity Supply Regulation 20.)

Typical CT ratings are given in BS 3938 and are 10, 15, 20, 30, 50 or 75 A in the primary with 1, 2 or 5 A in the secondary. Current transformers with bar primaries are made for circuits carrying several thousands of amperes however.

VOLTAGE OR POTENTIAL TRANSFORMERS

A d.c. ammeter is converted into a voltmeter by the addition of a series resistor or multiplier which limits the current at full rated voltage to that required to give full scale deflection of the movement. Up to about 1000 V this arrangement is quite satisfactory. Insulating the meter movement, the terminals and the multiplier from the case and the panel in which the meter is situated presents no special problems.

Above 1000 V hazards begin to present themselves. The cables to the meter may be long and are vulnerable to damage. Insulation becomes difficult and increasingly expensive as voltage rise.

Where alternating voltages are to be measured, the voltage transformer is used to reduce the voltage at the meter to 63.5 V or 110 V typically.

The voltage transformer is essentially a power transformer designed to minimise the core loss. Great care is taken to obtain maximum flux linkage between the coils and the winding resistance is made very small by using conductors with a larger cross-sectional area than in a power transformer of similar rating. The secondary is used for measuring purposes only so that the current is small. The internal volt drops may generally be ignored so that

$$\frac{V_p}{V_s} = \frac{N_p}{N_s} \text{ (Closely)}$$

Figure 5.19 shows a voltage transformer connected to one phase of an 11 kV system. The transformer has a ratio of 100:1. The dial of the voltmeter is marked to indicate the voltage on the high voltage side allowing for the transformer ratio.

Where voltage transformers are used to measure the line voltages on a three-phase system the secondaries are star connected giving line voltages of $63.5 \times \sqrt{3} = 110$ V at the meters.

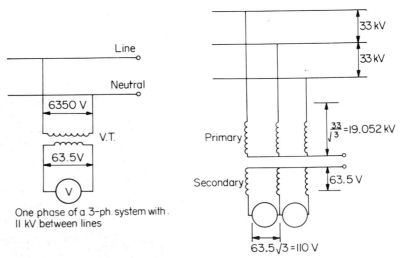

One phase of a 3-ph. system with 11 kV between lines

Figure 5.19   **Figure 5.20**

MEASUREMENT OF POWER AND POWER FACTOR

Current and voltage transformers are used to isolate wattmeters from the high voltage system in which the power is to be measured. The connections for a single-phase wattmeter are shown in *Figure 5.21*. The voltage and current coils are connected on one side to earth for safety reasons as already outlined.

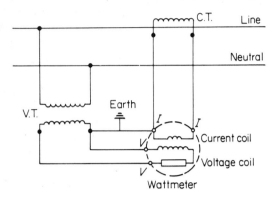

Figure 5.21

Example 5. A voltage transformer of ratio 100:1 and a current transformer of ratio 100:5 are used to measure the power and power factor in a single phase circuit using a wattmeter connected as shown in *Figure 5.21*. The potential difference across the wattmeter voltage coil is 63.5 V and the current in the current coil is 4.3 A. The wattmeter reading is 245 W.

Calculate for the primary circuit: (a) the current (b) the phase voltage (c) the power factor (d) the power.

(a) The CT ratio = 100:5

With 4.3 A in the secondary, the primary current
$$= 4.3 \times \frac{100}{5} = 86 \text{ A}.$$

(b) The voltage transformer ratio = 100:1

Primary phase voltage = $100 \times 63.5 = 6350$ V.

(c) The power factor in the secondary circuit is the same as that in the primary circuit assuming perfect transformers.

$$\text{Power factor} = \frac{\text{Power}}{\text{Volt-amperes}} = \frac{245}{63.5 \times 4.3} = 0.897.$$

(d) Power in the primary = $VI \cos \phi$ watts
$$= 6350 \times 86 \times 0.897$$
$$= 490\,000 \text{ W}.$$

This is the same as secondary power × VT ratio × CT ratio

$$= 245 \times \frac{100}{1} \times \frac{100}{5} = 490\,000 \text{ W}.$$

Example 6.   A load of 25 kVA with power factor 0.6 lagging is fed at 450 V from a single phase supply:

A voltage transformer with ratio 4:1 and a current transformer with ratio 100:5, together with a wattmeter, are used to measure the power in the circuit. Determine: (a) the potential difference across the wattmeter voltage coil; (b) the current flowing in the wattmeter current coil; (c) the indicated power on the wattmeter.

PROBLEMS FOR SECTION 5

7 What are the advantages of forming a transformer core from a continuous strip of cold rolled grain-oriented steel as compared with building it up from lamination stampings of the same material?

8 Explain what is meant by the term 'leakage flux'.

9 Why is the secondary of a three-phase distribution transformer star-connected?

10 What is the function of a conservator tank as fitted to a power transformer?

11 Describe three methods of cooling power transformers.

12 Why are some three-phase transformers with very large ratings auto-connected?

13 Give three advantages of the auto-connection for transformers.

14 Give two disadvantages of the auto-connection for transformers.

15 Estimate the ratio of copper in an auto-transformer to that in a double-wound transformer for ratios (a) 400 kV:132 kV (b) 400 kV:275 kV.

16 Explain the possible effects of open-circuiting the secondary of a current transformer whilst the primary is carrying current.

17 Explain why the secondary winding of a current transformer is earthed.

18 Explain why it is usual to employ current and voltage transformers when measuring the power in a high-voltage circuit.

19 A load of 20 kVA is fed at 240 V from a single-phase supply. A current transformer with ratio 80:5 and a voltage transformer with ratio 5:1 are used to drive a wattmeter which indicates 200 W. Determine: (i) the current in the wattmeter current coil, (ii) the voltage on the wattmeter voltage coil and (iii) the load power factor.

6 DC machines

Aims: At the end of this chapter you should be able to:

Describe the construction of a d.c. machine.
Describe the principle and construction of armature windings.
Describe the function of a commutator and the process of commu-
  tation.
Derive the e.m.f. equation of a d.c. machine.
Understand that a reversal of armature current is necessary for a
  reversal of energy flow so that $E = V \pm I_a R_a$.
Discuss the connection between magnetic flux, armature current
  and torque.
Describe the necessity for and the operation of a d.c. motor starter.
Describe the losses which occur in, and draw a power flow diagram
  for, a d.c. machine.

PRINCIPLE OF ACTION

From Faraday's original work we know that $e = d\Phi/dt$ volts for a single turn where e = electromotive force in volts, $d\Phi$ = flux linked in webers and dt = time interval in seconds.

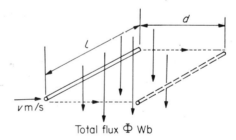

Total flux Φ Wb

Figure 6.1

Figure 6.1 shows a conductor of length l metres moving through an area of uniform magnetic field. The total flux is ΦWb and the conductor moves a distance of d metres at a uniform velocity of v m/s.

$$\text{Time taken} = \frac{\text{distance}}{\text{velocity}} = \frac{d}{v} \, s$$

Total flux linked = flux density × area of the field.

$$= B \times l \times d \text{ webers}$$

$$e = \frac{\text{flux linked}}{\text{time}} = \frac{B \times l \times d}{d/v} = Blv \text{ volts.}$$

If the ends of the conductor are joined to an external circuit a current i amperes will flow.

Power = ei watts   Work done = power × time = ei × time

Therefore, work done $= \left(\dfrac{\text{flux linked}}{\text{time}} \right) \times i \times \text{time} - Bldi$ joules.

But work done = force × distance operated through

Therefore, $Bldi$ = force × d

Force = Bli newtons.

Thus to generate an e.m.f. the requirement is for magnetic flux, a conductor and relative motion between them. To produce a force on a conductor the requirement is for magnetic flux and a current flowing in that conductor.

In generators the flux is usually provided by a system of electro-magnets and the motion of the conductors produced using an engine.

In motors the current is provided from an external source. This current both excites the electromagnets and feeds the conductor system.

In the d.c. motor the force created by the current flowing in the conductor system causes them to accelerate if they are free to move. The motion of the conductors through the field causes an e.m.f. to be generated. By Lenz's law, the generated e.m.f. opposes the supply voltage. For this reason it is called the back e.m.f. At a certain speed dependent on the mechanical forces involved, the strength of the magnetic field and the applied voltage, equilibrium will be reached between the applied voltage and the back e.m.f.

The applied voltage is given the symbol V
The back e.m.f. is given the symbol E.

In order that current shall flow into the motor to produce the necessary force and torque to sustain rotation, E must always be less than V.

$V - E = IR$ where the conductor current = I amperes and the conductor resistance = R ohms.

If the mechanical load on the motor is increased it will slow down so reducing the velocity of the conductors through the field. This reduces the back e.m.f. so causing the current to increase. The increase in current provides the extra force necessary to keep the motor turning against the increased load.

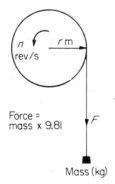

Figure 6.2

POWER, FORCE AND TORQUE

Figure 6.2 shows a pulley connected to a motor. It is arranged to wind up a rope, the downward force in which is F newtons. The force is provided by a suspended mass. In one revolution of the pulley the mass is raised through $2\pi r$ metres.

Work done = Force × Distance = $F \times 2\pi r$ newton metres (joules)

The pulley rotates at n rev/s

Therefore, work done per second = $F \times 2\pi r \times n$ J/s

Now 1 J/s = 1 watt and $F \times r$ = torque T

Hence power = $2\pi n T$ W

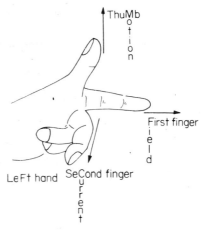

Figure 6.3 Fleming's left hand rule

FLEMING'S RULES

For motors   The relationship between the direction of the force, current and magnetic field has been determined by experiment and a convenient way of remembering this is by the use of Fleming's left hand rule.

In *Figure 6.3* the le*F*t hand relates current and magnetic field to *F*orce and force is produced by an electric motor to drive its load. The thu*M*b indicates the direction of the force and hence *M*otion
The *F*irst finger represents the direction of the *F*ield
The se*C*ond finger represents the direction of the *C*urrent

For generators   For *G*enerators the ri*G*ht hand is used. The thumb and two fingers represent the same quantities as for motors. In a generator the conductor system is driven by an external force and a current flows such as to produce a force which opposes that motion. This is clear from *Figures 6.4* and *6.5*. For the generator, the motion and field are unchanged but the current direction has reversed.

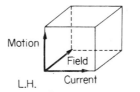

Figure 6.4

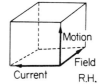

Figure 6.5

THE ACTION OF THE COMMUTATOR   Consider now the very simple generators shown in *Figure 6.6*. In both cases the magnetic fields are stationary and the coil rotates. *Figure 6.6(a)* is an alternating current generator and the output is taken from the coil using conducting sliprings on which blocks of carbon called brushes rub. Early machines used wire gauze made up like a sweeping brush and the name is still used. With the coil in the position shown the generated e.m.f. is zero since the coil sides are not cutting the lines of flux. As the coil revolves the voltage increases

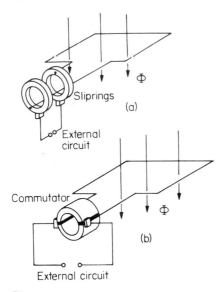

Figure 6.6

to a maximum after a rotation of 90° when the coil sides are moving directly across the flux. After a further 90° rotation the voltage will again be zero. As the rotation continues a voltage of reverse polarity will be produced, rising to a maximum and falling to zero as the revolution is completed since the direction of motion of the coil sides is reversed during this period. A coil side which moved from right to left during the first half revolution moves from left to right during the second half revolution.

Now consider the same coil but this time connected to a two segment commutator as shown in *Figure 6.6(b)*. This simple commutator is a split copper cylinder, each half fully insulated from its neighbour. As before, carbon brushes are used to connect the external circuit.

The operation of the commutator can be understood by looking firstly at *Figure 6.7* The brushes are shown inside the commutator for clarity. Assume that there is an external load connected so that current flows in the coil. The left hand coil side connected to commutator segment 1 (hatched) has the current direction shown. This is deduced using Fleming's right hand rule. Current is flowing in the coil towards the left hand brush which delivers current to the external circuit and therefore has positive polarity. Current returns from the circuit to the right hand brush and flows in the coil away from the commutator segment.

In *Figure 6.8* we see the coil turned through 90°. The coil sides are outside the field so that the generated e.m.f. is zero. The brushes are shorting out the coil since they are touching both halves of the commutator at the same time. Later, when commutators with many more segments are considered it must be remembered that the change over of connection from one segment to the next must occur when the associated coil sides are out of the magnetic field.

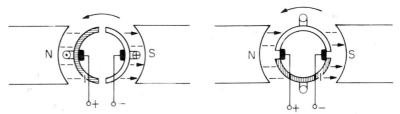

Figure 6.7   Figure 6.8

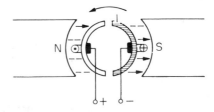

Figure 6.9

In *Figure 6.9* the coil has turned through a further 90° and commutator segment 1 is now on the right. The current direction in the coil side connected to it has reversed but since a commutator is being used the left hand brush is now connected to the other side of the coil and so still has the same polarity.

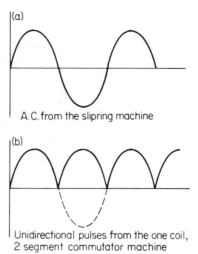

Figure 6.10

The e.m.f. generated in the coil is alternating but by using a mechanical reversing switch or commutator, the current flowing in the external circuit is a series of unidirectional pulses.

THE RING-WOUND ARMATURE

In *Figure 6.11* the laminated iron core is wound with ten coils each having two turns. The commutator has ten segments, one for each coil. The magnetic flux produced by the poles crosses the air gap into the core. Only the outsides of the turns directly under the poles cut the flux and the directions of the induced e.m.f.s shown are again deduced using Fleming's right hand rule. In both the top and bottom halves of the winding this direction is towards the right hand brush which is therefore positive.

With the armature in the position shown each brush is shorting out one coil which has no e.m.f. induced in it since it is not cutting the flux. The flux at this instant is passing through the coil which is the required condition for commutation as already described (see *Figure 6.8*). Each coil passes through this position in turn as the armature rotates and during this instant carries no current.

There are six conductors cutting the magnetic flux from each pole at any instant in the arrangement shown in *Figure 6.11*. Representing the e.m.f. induced in each conductor by a cell, each half of the winding has six cells in series. Both halves of the winding are in parallel.

When a load is connected each half of the winding carries one half of the load current. As with the single coil, the direction of the e.m.f. induced in individual coils is alternating being in one direction as the coil passes under the north pole and in the opposite direction as it passes under the south pole. In this machine the air gap between the pole faces and the iron core is of uniform length so that the flux is uniform in the gap.

As a coil passes under a pole from point A to point B in *Figure 6.13*, the induced voltage is nearly constant. While moving from point B to point C the voltage falls to zero since no flux is being cut. The waveform of the voltage in a single coil is shown in *Figure 6.14*.

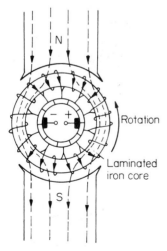

Figure 6.11

number 1 and its two sides are labelled 1 and 1′ respectively. Its
t w o

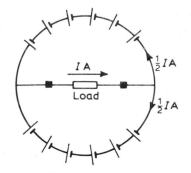

Figure 6.12

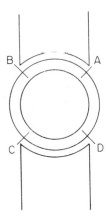

Figure 6.13

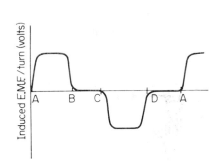

Figure 6.14

This should be compared with that shown in *Figure 6.10* which
was for a different coil and field configuration.

There are two disadvantages to the ring winding. These are:

1   The right winding is difficult to wind since each turn must be
taken round the core by hand.

2   Only a small part of the winding is active at any one time.

Other windings have been developed to overcome these problems.

THE LAP WINDING

Each coil in the lap winding overlaps its neighbours.

Figure 6.15 shows a lap coil comprising two turns. It is coil
number 1 and its two sides are labelled 1 and 1′ respectively. Its two
ends are connected to adjacent segments on a commutator and coil
number 1 starts on segment number 1. The other end of the coil is
connected to segment number 2.

Coil number 2 is added and this starts on segment number 2 of the
commutator. Only one turn per coil is shown here making the diagram easier to follow. There may in fact be many turns in each coil.

The coils are situated in slots in a laminated iron armature and
Figure 6.17(a) shows a part wound armature. *Figure 6.17(b)* shows
how the windings overlap. The complete winding is built up in this
manner until all the slots in the armature contain two coil sides, one
on top which starts at the commutator segment which carries its
number, 1, 2, 3 etc. and the other at the bottom which is the return
coil side numbered 1′, 2′, 3′ etc. The fully wound armature rotates
in a magnetic field which is generally produced electrically. A
typical four-pole arrangement is shown in *Figure 6.18*.

Consider the typical section of a lap winding under a north and
south pole as shown in *Figure 6.19*. The view is from the centre of
the armature looking outward through the winding and seeing the
pole faces outside the winding. The lines of force from the north
pole are straight up out of the paper and those from the south pole
down into the paper. Using Fleming's right hand rule the directions
of the induced e.m.f.s are as shown. Between the positive and negative brushes there are six conductors connected in series so that the
e.m.f. between the brushes is six times that induced in a single
conductor. The complete lap winding has one brush for each pole

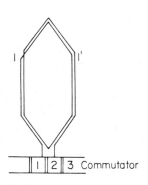

Figure 6.15

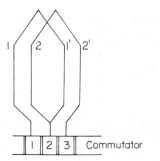

Figure 6.16

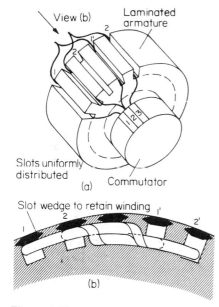

Figure 6.17

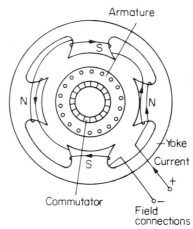

Figure 6.18

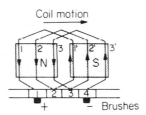

Figure 6.19

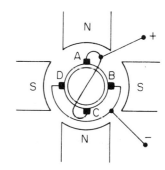

Figure 6.20

and the brushes under like poles are connected together. For a four-pole machine there are four brushes and the arrangement is shown in *Figure 6.20*.

The winding shown in *Figure 6.19* is repeated four times between brushes A and B, B and C, C and D and D and A. The full winding for this simple machine therefore comprises 24 conductors (12 turns), conductor 12′ returning to segment 1 under conductor 3 to complete the winding.

There are therefore four parallel paths through the armature:

From B to A, B to C, D to A and D to C.
Each path comprises six conductors in series.

Example 1. A 6-pole d.c. lap wound generator has 36 slots on the armature. Each slot contains 8 conductors. When the speed of the armature is 500 rev/min, the induced e.m.f. in each conductor is 5 V.

Calculate: (a) The total number of conductors on the armature.
(b) The total number of turns on the armature.
(c) The number of conductors in series in each of the parallel groups.
(d) The total generated e.m.f.

(e) The rating of the generator if each conductor can carry 10 A before overheating.

(a) Total number of conductors = number of slots × number of conductors in each slot. = 36 × 8
$$= 288.$$

(b) Two conductors form one turn. Number of turns = $\dfrac{288}{2}$
$$= 144.$$

(c) Since there are six poles, there will be six brushes and six parallel paths through the armature.

Number of conductors in series in each group = $\dfrac{288}{6}$ = 48.

(d) Generated e.m.f. = e.m.f. per conductor × number of conductors connected in series.
$$= 5 \times 48$$
$$= 240\,\text{V}.$$

(e) Each conductor can carry 10 A and there are six parallel paths.

Total current from the armature = 6 × 10
$$= 60\,\text{A}.$$

Rating = 240 × 60
$$= 14\,440\,\text{W}.$$

THE WAVE WINDING

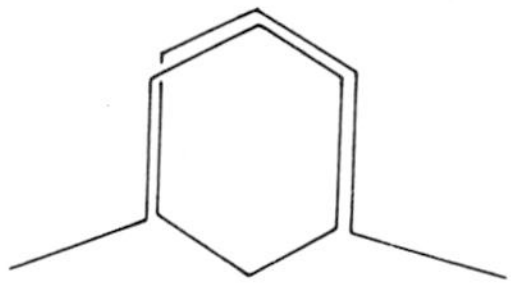

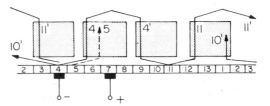

Figure 6.21

Figure 6.21 shows a coil for a wave winding. The ends of the coil are not connected to adjacent segments on the commutator but to segments some distance apart. Again using a four-pole machine as an example, the winding is shown in part in *Figure 6.22*.

Notice that commutator segments 2 and 3 are repeated at each end so that the commutator may be visualised in its circular form. There is an additional commutator segment, number 13, which causes the winding to progress in 'waves'.

Starting with the negative brush presently on segment 4, follow conductors 4 and 4′ to segment 11, and then conductors 11 and 11′ to segment 5. Four conductors are involved in progressing one segment along the commutator. The process may be continued following four more conductors (not shown) so returning to segment 6. A further four conductors, making twelve in all allows us to progress to segment 7 upon which the positive brush rests.

Figure 6.22

With this winding two parallel paths through the armature exist. One of them is as described and the other could be traced from segment 4 once again starting with conductor 10′ which passes under the south pole on the right of the diagram, returning again to segment 7 eventually.

A wave wound machine has only two brushes and two parallel paths through the armature irrespective of the number of poles.

Example 2. A four-pole wave wound generator has the following details:
Number of slots on the armature 64
Conductors per slot 15
Induced e.m.f. per conductor 1.8 V
Maximum current per conductor 15 A
Determine: (a) the output voltage (b) the maximum current which the machine may safely deliver.
(a) Total number of conductors = 15 × 64 = 960
There are two parallel paths, hence number of conductors in series per path = $\frac{960}{2}$ = 480.

Total induced e.m.f. = 1.8 × 480 = 864 V.
(b) Maximum current = 2 × 15 = 30 A.

Example 3. A d.c. generator has 4 poles and 72 conductors on the armature. The e.m.f. generated per armature conductor at a particular speed is 10 V. The current carrying capacity of each conductor is 20 A. Calculate the output voltage and maximum power rating for (a) lap connection (b) wave connection.

THE EMF AND SPEED EQUATIONS

Figure 6.23 shows part of a d.c. machine. Consider a single conductor on the armature as it moves from position A to position B.
Let Φ = total flux per pole in webers
n = speed of the armature in revolutions per second.
P = number of *pairs* of poles on the yoke. (Pairs are used since it is not possible to have a single pole.)
Z = total number of conductors on the armature
c = number of parallel paths through the armature.
There are 2P poles on the machine and at n rev/s the particular conductor shown will therefore pass 2Pn poles every second.

The time taken to pass one pole = $\frac{1}{2Pn}$ s

This is the time taken for the conductor to move from A to B in *Figure 6.23*. During this time the flux Φ from one pole is cut.

For a single conductor $e = \dfrac{\text{flux cut}}{\text{time (s)}}$ volts.

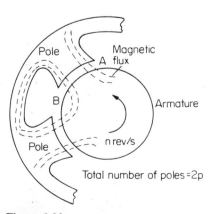

Figure 6.23

Hence the average voltage induced in the conductor $= \Phi / \dfrac{1}{2Pn}$

$$= 2Pn\,\Phi \text{ volts}$$

On the armature there are Z/c conductors in series (see Examples 1 and 2).

Hence, the total e.m.f. $E = 2Pn\,\Phi\,\dfrac{Z}{c}$ V

Transposing: $n = \dfrac{E}{2P\,\Phi\dfrac{Z}{c}} = \dfrac{Ec}{2P\,\Phi\,Z}$ rev/s

Since 1 revolution $= 2\pi$ radians

$$\omega = \frac{2\pi Ec}{2P\,\Phi\,Z} = \frac{\pi Ec}{P\,\Phi\,Z} \text{ rad/s}$$

Example 4. The wave wound armature of a six-pole d.c. generator has 30 slots and in each slot there are 8 conductors. The flux per pole is 0.0174 Wb. Calculate the value of the e.m.f. generated when the speed of the armature is 1200 rev/min.

$$1200 \text{ rev/min} = \frac{1200}{60} = 20 \text{ rev/s}$$

For a wave wound armature $c = 2$
6 poles $= 3$ pole pairs. Hence $P = 3$.

$$E = 2 \times 3 \times 20 \times 0.0174 \times \frac{30 \times 8}{2} = 250.6 \text{ V}$$

Example 5. A lap wound d.c. generator is to have an output voltage of 500 V at 26 rev/s. The armature has 28 slots each containing 12 conductors. Calculate the required value of flux per pole.

For a lap winding there are the same number of parallel paths through the armature as there are poles.

There $c = 2P$ and $E = \Phi nZ$ volts.
$$500 = \Phi \times 26 \times 28 \times 12$$

$$\Phi = \frac{500}{26 \times 28 \times 12}$$

$$= 0.057 \text{ Wb}.$$

DC MACHINE CONNECTIONS

Shunt *Figure 6.24* shows the shunt connection. The field winding comprises many turns of fairly fine wire and it is connected in series with

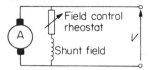

Figure 6.24

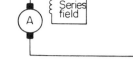

Figure 6.25

a control rheostat directly across the supply. It is therefore in parallel with, or shunting, the armature.

The field current is independent of the armature current unless this is large enough to cause a voltage drop in the supply lines in which case both the armature voltage and the field voltage will be affected.

Series *Figure 6.25* shows the series connection. The field winding comprises a few turns of very heavy gauge wire or copper strip. It is connected in series with the armature and so carries the same current. For this reason it must have a very low resistance or the power loss will be excessive.

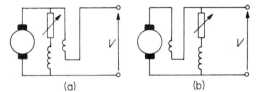

(a)   (b)

Figure 6.26

Compound *Figures 6.26(a)* and *(b)* shows machines with some shunt field and some series field. Both shunt and series coils are wound on the same pole pieces and the series field may either assist or oppose the shunt field. In the former case the machine is said to be 'Cumulatively wound' and in the latter case 'Differentially wound'. *Figure 6.26(a)* shows what is known as the short shunt connection whilst *Figure 6.26(b)* shows a long shunt.

THE DC MOTOR

Speed/armature current characteristics When a voltage V is applied to the armature of a motor a current I_a flows. A force is produced which causes the armature to accelerate and a back e.m.f. E volts is generated. Clearly V must always be larger than E since if it were not so then current would not flow into the armature and no work could be done. The armature resistance $= R_a \Omega$.

Equilibrium is reached when $E = V - I_a R_a$ (6.1)

Transposing: $I_a R_a = V - E$ Therefore $I_a = \dfrac{V - E}{R_a}$

Now $n = \dfrac{Ec}{2P\phi Z}$ rev/s.

For a particular machine P, Z and c are constants so let $\dfrac{2PZ}{c} = k$, a constant.

Now $n = \dfrac{E}{k\Phi}$ (6.2)

From equation 7.1 we see that any increase in I_a will cause the voltage drop $I_a R_a$ to increase and the voltage E to decrease.

In a shunt wound motor, if the field rheostat is not adjusted, the flux Φ will remain constant and from equation 6.2 since both k and Φ are constant, E and n are directly proportional.

In a series motor the field winding carries the armature current. The resistance of the armature circuit must include that of the series field since it will cause a voltage drop in addition to that of the armature itself.

Hence $E = V - I_a\,(R_a + R_{sf})$ where R_{sf} = resistance of the series field.

Ignoring the effect of magnetic saturation, the flux set up by the field winding is proportional to the armature current.

Again $n = \dfrac{E}{k\Phi}$ (6.2)

Now since increasing the armature current increases the flux, for a constant value of E the effect of increasing the load (and I_a) is to reduce the speed. Since in fact E falls as the armature current is increased (equation 6.1) there is a further reduction in speed due to this.

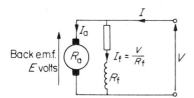

R_a = resistance of the armature
R_f = resistance of the shunt field
$I = I_a + I_f$

Figure 6.27

Example 6. A d.c. shunt motor has an armature circuit resistance of $0.5\,\Omega$ and a shunt field resistance of $240\,\Omega$. It is connected to a 240 V supply. On no-load the input current is 2 A and the speed 1500 rev/min.

When fully loaded the input current is 8 A. Calculate the value of back e.m.f. generated and the speed of the motor at full load.

$$I_f = \frac{VI}{RF} = \frac{240V}{240\,\Omega} = 1\,A$$

1500 rev/min = 25 rev/s

$I = I_a + I_f$ (from *Figure 6.27*)   Therefore $I_a = 2 - 1 = 1$ A on no load.

$E_1 = V - I_a R_a$ (Using suffix 1 to indicate original no-load conditions)

$\quad = 240 - 1 \times 0.5$
$\quad = 239.5$ V

$n_1 = \dfrac{E_1}{k\Phi}$ Transposing: $k\Phi = \dfrac{239.5}{25} = 9.58.$

When the total current = 8 A, $I_a = 7$ A (The field current remains constant)

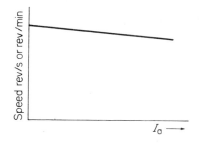

Figure 6.28

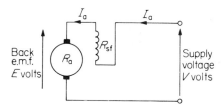

Figure 6.29

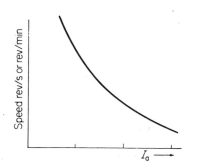

Figure 6.30

$E_2 = 240 - 7 \times 0.5$ (Suffix 2 indicates the new fully loaded conditions)

$$= 236.5 \text{ V}$$

$$n_2 = \frac{E_2}{k\Phi} = \frac{236.5}{9.58} = 24.68 \text{ rev/s}$$

Figure 6.28 shows a typical speed/armature current characteristic for a shunt motor.

Example 7. A series motor has an armature resistance of $0.15 \, \Omega$ and a series field resistance of $0.25 \, \Omega$. It is connected to a 250 V supply and at a particular load runs at 30 rev/s when drawing 10 A from the supply. Calculate the speed of the motor when the load is changed such that the armature current is increased to 20 A.

$$E_1 = V - I_a (R_a + R_{sf})$$
$$= 250 - 10(0.15 + 0.25) = 246 \text{ V}$$

As in Example 6, $k\Phi_1 = \dfrac{E_1}{n_1} = \dfrac{246}{30} = 8.2$

When $I_a = 20 \text{ A}$, $E_2 = 250 - 20(0.15 + 0.25) = 242 \text{ V}$

Now since the armature current has doubled, the flux has doubled.

$$n_2 = \frac{E_2}{k\Phi_2} \text{ where } k\Phi_2 = 2k\Phi_1 = 16.4$$

$$n_2 = \frac{242}{16.4} = 14.93 \text{ rev/s}$$

Thus as armature current increases the speed falls. *Figure 6.30* shows a typical speed armature current characteristic for a series motor. Notice that for low armature currents the speed is high since the flux is low and for this reason the series motor must always be connected to a load which will limit its top speed. On no load the speed would be sufficiently high to create disruptive centrifugal forces and the commutator segments and windings would be thrown outwards.

Example 8. A lap wound armature for a four-pole d.c. machine has 56 slots each containing 10 conductors. For a flux per pole of 36 mWb, calculate: (a) the generated voltage E at a speed of 25 rev/s, (b) the speed at which the machine will run as a motor when drawing an armature current of 25 A from a 600 V supply given that the armature resistance is $0.5 \, \Omega$.

Example 9. Repeat question (3) assuming that the armature is wave wound, all other conditions remaining unchanged.

A level characteristic may be produced by compounding. With the shunt motor the speed falls as the load is increased. By adding a small differentially wound series winding to the main poles the total flux per pole is reduced as the load current increases. Since speed is inversely proportional to flux, this causes an increase in speed compensating for the reduction due to the resistance of the armature.

Cumulative windings may be used to give a motor a characteristic between that of the series and shunt motors. The speed will fall as the load is increased but the motor will have a safe maximum speed on no load due to the constant shunt field.

The torque equation

From previous work we know that for a shunt motor

$$E = V - I_a R_a$$

Multiply throughout by I_a

$$EI_a = VI_a - I_a^2 R_a \quad \text{(all watts)}$$

VI_a = total power input to the armature (supply voltage × current)

$I_a^2 R_a$ = power loss in the armature as heat.

Hence EI_a must be the power available at the armature shaft to produce torque. (Note that by using $(R_a + R_{sf})$ throughout, the proof is valid for the series motor.)

Now $E = \dfrac{2P\Phi Zn}{c}$ volts and power = $2\pi n T$ watts

Therefore $EI_a = \dfrac{2P\Phi Zn}{c} I_a = 2\pi n T$

Transposing $T = \dfrac{2P\Phi Zn}{c2\pi n} I_a = \dfrac{P\Phi Z}{c\pi} I_a$ newton metres.

As before let $\dfrac{2PZ}{c} = k$

When $T = \dfrac{k}{2\pi} \Phi I_a$ Nm.   Where T = gross torque.

The gross torque developed is proportional to the flux and to the armature current. Not all this torque is available to drive an external load however since there will be friction in the motor bearings and between the armature surface and the air in the casing. The armature may be driving a fan which forces air over the windings to keep them cool and a torque is required to drive this. Finally, since the iron core of the armature is being driven through a magnetic field there will be hysteresis and eddy current losses.

Gross torque − all loss torques = net torque available to drive an external load.

Torque/armature current characteristics

In the shunt motor, for a constant rheostat setting, the flux Φ is constant.

Now $T = \dfrac{k}{2\pi} \Phi I_a$ Nm

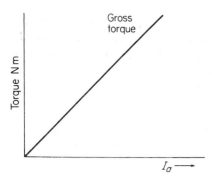

Figure 6.31

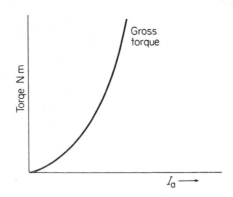

Figure 6.32

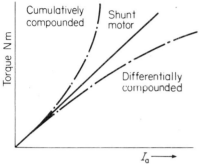

Figure 6.33

Therefore I_a is the only variable and the gross torque is directly proportional to the armature current. This is shown in *Figure 6.31*.

In the series motor the field is produced by the armature current. $\Phi = k_1 I_a$ up to saturation where k_1 is the constant of proportionality between flux and current in webers per ampere. Substituting in the torque equation:

$$T = \frac{k}{2\pi} k_1 I_a^2$$

which is the equation of a parabola $[y = mx^2]$. This is shown in *Figure 6.32*.

Differential compounding will cause the torque produced to be less than that of the shunt motor for a given value of armature current whilst cumulative compounding will increase the available torque.

Motor applications

Shunt motors are used where virtually constant speed is required on drives such as machine tools, fans and conveyor systems.

Series motors produce very high torques at low speeds and so they are suitable for starting very heavy loads. The main uses are in traction being used extensively in trains, electric buses, trams and milk delivery vehicles.

Compound wound motors are used where constant speed is required (differential winding) for coal feeders on boilers, and oil pumps, or where a high torque on starting and safe top speed are required (cumulative winding) for conveyor systems, hoists and cranes.

Speed control

A measure of speed control may be achieved by changing the field current of the motor. Since the induced voltage E is proportional to flux and speed any change in magnetic flux will result in a change in speed. The lower is the flux the higher is the speed. However torque is proportional to flux and armature current so that as the flux is reduced the armature current must increase if the load torque remains constant as the speed changes. This is often the limiting factor since large armature currents will cause overheating.

Usual values of speed obtainable by field weakening on a standard motor will be from normal up to something less than twice

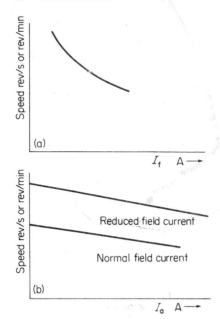

Figure 6.34

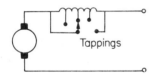

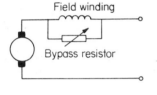

Figure 6.35

normal although a larger speed range is achievable using specially designed motors.

Since steel saturates at less than 2 tesla, increasing the field current and hence field strength as a method of speed reduction has very limited applications. This must be done by reducing the voltage on the armature. If the field and torque remain constant, the armature current and volt drop will remain constant.

Since $E = V - I_aR_a$, as the applied voltage V is reduced so the speed will fall in proportion in order to satisfy the e.m.f. equation.

Figure 6.34 shows characteristics of a shunt motor employing varying field currents.

In shunt motors a reduction in field current is achieved by using the field control rheostat (*Figure 6.24*).

In series motors a reduction in flux may be achieved by either by-passing part of the armature current round the field winding or by having a tapped field winding which allows a variable number of turns to be used. These two methods are shown in *Figure 6.35*.

Reduction of armature voltage can be obtained by using an additional resistor in the armature circuit but this is extremely wasteful of power since the whole of the armature current flowing in this resistor produces a considerable amount of heat. If the speed is reduced to one half normal by this method then one half of the input power is wasted.

Where large motors are involved controlled rectifiers or a d.c. generator may be used to provide the varying voltage. Silicon controlled rectifiers are often used when the mean output voltage can be controlled from zero to its full rated value at very high efficiency.

Where a d.c. generator is used, this is driven at constant speed, often by an a.c. motor. The output voltage from the generator may be varied from zero to full rated value by increasing its field current. By reversing the field connections the polarity of the output voltage is reversed. A motor supplied from this generator can be made to run at any speed from a crawl to its full rated value in either direction. Such a system is named after Ward-Leonard, the developer.

DC MOTOR STARTERS

Since $E = \dfrac{2P\Phi nZ}{c}$ *volts*

it follows that when a motor is at rest, since n = zero, whatever value of flux, the back e.m.f. is zero.

But $E = V - I_aR_a$ Therefore $0 = V - I_aR_a$

Transposing $V = I_aR_a$ or $I_a = \dfrac{V}{R_a}$ amperes.

The resistance of the armature is kept as small as possible to keep the losses in the armature to a minimum. It follows that the armature current under these conditions will be extremely large. It could in fact be large enough to cause severe damage to the armature by heating the conductors and commutator. Also since there are mechanical forces between current carrying conductors, with large currents sufficient force may be developed to disrupt the winding.

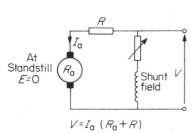

$V = I_a(R_a + R)$

Figure 6.36

Measures must be taken to prevent these excessive currents flowing. Where the motor is supplied from direct mains this may be achieved either by the addition of resistance to the armature circuit or by electronic means employing a 'chopper' circuit.

Where the motor is fed from an alternating supply, controlled rectifiers will be used to vary the level of input voltage and hence the starting current.

The addition of resistance to the armature circuit

In *Figure 6.36* the resistance R ohms is included in the armature circuit so that at standstill

$$0 = V - I_a(R_a + R) \text{ or transposing: } I_a = \frac{V}{(R_a + R)}$$

The added resistance is made large enough to limit the starting current to between 1.5 and 2 times the normal full load current of the motor. As the motor accelerates the back e.m.f. rises from zero so that

$$I_a = \frac{V - E}{(R_a + R)}.$$

The value of the added resistance can now be decreased somewhat to allow the armature current to rise to its original value once more. Further acceleration and an increase in the back e.m.f. will take place.

Reductions in additional resistance are made at intervals as the speed of the motor builds up. Finally when the armature resistance alone remains

$$I_a = \frac{V - E}{R_a}$$

During the starting process it is essential to have the maximum possible field strength in the motor since both torque and back e.m.f. are proportional to the field flux. The additional resistance must therefore only affect the armature circuit whilst the field winding is connected directly across the supply.

The manually-operated face-plate starter

Figure 6.37 shows the armature circuit only of a d.c. shunt motor with a suitable starter. The handle is moved manually from the off position to make contact with the first resistance stud. This puts

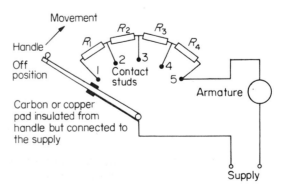

Figure 6.37

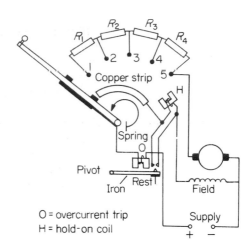

Figure 6.38

resistances R_1, R_2, R_3 and R_4 in series with the armature. As the motor speed rises from zero to a few revolutions per second the handle is moved so that the supply is fed to contact stud number 2. Resistances R_2, R_3 and R_4 are now in series with the armature. As the speed builds up the handle is moved successively across the studs ending up on number 5 when the armature is connected directly to the supply. As stated previously, the field current must be a maximum during this period and a method of achieving this is shown in *Figure 6.38*.

The starting handle being moved on to the first stud puts all the resistances in series with the armature. The current from the supply positive terminal flows through O, the overcurrent trip to the bottom of the starting handle along which there is a conducting strip to two contact pads. One of these feeds current to the series resistors whilst the other makes contact with the copper strip which is connected to the field winding. On the way to the field winding the current flows through H, the hold-on coil. The field winding is therefore subjected to the full supply voltage (less a very small voltage drop in coil H). As the starting handle progresses to the running position the field remains unchanged.

When the handle reaches stud 5 it comes into contact with the faces of the electromagnet H. Either the whole handle or just a small pad fixed to it is made of a magnetic material so provided that a suitable current is flowing in the field winding, the handle will be held in the running position.

All the current being supplied to the motor flows through the small coil on the horseshoe shaped piece of iron of the overcurrent trip O. A beam beneath the coil is held on the rest stop by gravity or a small spring. Normally the magnetic effect of the current is too small to attract the beam upwards. If however the current becomes excessive the beam is pulled upwards and the two electrical contacts to the right are made which causes the field current to be bypassed round the hold-on coil. This becomes de-energised and the starting handle is released. A spring causes it to fly back to the start position.

If at any time the supply to the motor is lost, the coil *H* will be de-energised and the handle will return to the start position. This prevents the motor from starting up unexpectedly when the supply is restored with possibly disastrous results both to the motor and to the operator who may have a hand in the driven equipment.

The automatic contactor-type starter

The manually operated starter suffers from a number of disadvantages amongst which are (i) there will always be some burning of the contacts, especially in the larger sizes, as the operator moves the handle across the face of the starter and (ii) there is no real control over the speed at which the handle is moved (apart from possibly the overload trip which will cut off the supply if the input current reaches a very high value). Better types of starter involve the use of contactors as shown in *Figure 6.39*. The motor is started by closing the main contactor. All the resistance is in circuit, as in the case of the manual starter. The motor speeds up when contactor C_1 is closed. This shorts out R_1, allowing the motor to gain more speed. Progressively the contactors are closed until finally, by closing C_5, the armature is connected directly across the supply. The field has to be connected to the full voltage during the run-up period, but this is not shown in the figure.

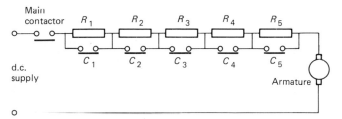

Figure 6.39

The contactors may be closed by using relays actuated:
(i) by the build-up of armature-generated voltage *E*. For example, with a motor being supplied from 400 V mains, when *E* reaches 50 V, C_1 closes, when *E* reaches 100 V, C_2 closes etc;
(ii) by the swings in armature current. When the motor is started the initial input current will be of the order of twice the full-load value. As the speed increases, the current value will fall. When it reaches 1.5 times full load current, C_1 closes, causing the current to rise and the speed to increase. When the current again falls to 1.5 times its full load value, C_2 closes and so on until all the contactors are closed.
(iii) on a time basis. Several run-ups are performed manually and the times taken to reach certain key speeds noted. These values are fed into a timing circuit which will close the contactors after these set periods during each run-up.

Electronic means

There is a considerable power loss when using the resistance type of starting. These losses may be avoided and often a much smoother run-up achieved using electronic starters. We will consider starting when employing (a) a direct supply and (b) an alternating supply.

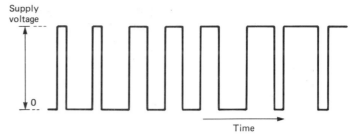

Figure 6.40

(a) *Direct supply.* The electronic chopper switches the mains supply directly across the armature for a very short period of time, typically a few milli-seconds. The current begins to rise towards a very high value but before this is achieved the supply is disconnected. After a further very short period the supply is re-connected and again disconnected. This switching takes place many times per second and over a number of seconds. The motor will be gathering speed all the while and the 'on' periods will be made progressively longer. This is shown in *Figure 6.40* in condensed form. There will be hundreds, if not thousands, of switching operations before the chopper circuit ceases to operate and the armature is connected directly to the supply. It is possible to hear the chopper at work on vehicles such as milk floats as a note that varies as it gathers speed.

(b) *Alternating supply.* A controlled rectifier employing thyristors is used to vary the voltage applied to the armature. An electronic circuit is used to cause the rectifier to become conducting at any particular point in the alternating cycle. Delaying conduction for most of the cycle means that the average voltage supplied to the armature is low. Reducing the delay means that a higher average voltage is delivered. In *Figure 6.41* a single-phase half-wave rectifier is considered for ease of explanation. During the first positive half-cycle the rectifier is switched on at t_1. The voltage delivered to the motor has a peak value p_1. The average voltage over the half-cycle is low so that the input current to the motor is not excessive. After many half-cycles, during which the motor is speeding up, the delay on the rectifier is changed to t_2. The peak voltage applied to the armature rises to p_2 and the average voltage is somewhat higher. After a suitable interval, the delay is further reduced and the figure

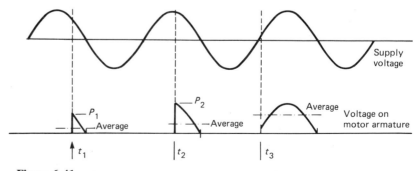

Figure 6.41

shows the instant just before the complete wave is used, t_3. The average voltage is now much higher. Generally a full-wave rectifier giving twice the number of pulses or a three-phase bridge rectifier giving six times the number of pulses illustrated will be used.

THE DC GENERATOR

Shunt-wound

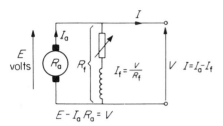

$E - I_a R_a = V$

Figure 6.42

If a d.c. machine connected to a supply and running as a motor, has the speed of its armature raised above that at which it operates as a motor by driving it with an external engine whilst keeping the field constant, the value of the generated e.m.f. will increase since this is proportional to speed.

As E becomes greater than the supply voltage a current will flow from the machine which has now become a generator, receiving power from the driving engine and delivering power to the electrical system. The difference between E and V is still $I_a R_a$.

$$E = V + I_a R_a.$$

For a motor $E = V - I_a R_a$ and the sign change for the generating condition indicates a change in the direction of the current. In the self-excited generator in *Figure 6.42* part of the armature output is used to excite its own field so that $I = I_a - I_f$. An increase in load current causes $I_a R_a$ to increase and the terminal voltage V will fall.

The field current $I_f = V/R_f$ so that a reduction in terminal voltage results in a reduction in the field current and hence the field flux.

Since $E = k\Phi n$ volts, the reduction in flux causes a reduction in generated e.m.f. The terminal voltage of a self-excited generator therefore falls quite rapidly as the load current increases due to the combined effects. A typical characteristic is shown in *Figure 6.43*

When driven at constant speed with no load connected, the relationship between field current and generated e.m.f. may be determined using the circuit shown in *Figure 6.42* with the addition of a voltmeter to measure output voltage and an ammeter in the field circuit. Due to magnetic saturation the output voltage is not proportional to the field current. With zero field current there is a small output voltage which is produced by the residual magnetism in the pole pieces without which the generator cannot commence generation. In a new machine this is created using an external power source. *Figure 6.44* shows the open circuit characteristic.

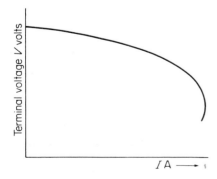

Figure 6.43

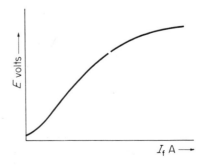

Figure 6.44

Series-wound

This is rarely if ever used since with no load on the machine there is no current in the field winding and the output voltage is near zero. When a current flows a flux is produced and the output voltage rises. The terminal voltage is therefore a function of the load current.

Compound-wound

Cumulative compounding is used to overcome the drooping voltage characteristic shown in *Figure 6.43* for the shunt machine. As load current flows in the series winding a flux is produced which is added to that of the shunt field. This increase in total flux produces a corresponding increase in generated e.m.f. The output voltage can thus be held constant or made to increase slightly to compensate for the voltage drop in the connecting cables to the external load.

Differential compounding is used to cause the terminal voltage to collapse when load current flows. Load current increases the series field which opposes the shunt field so causing a reduction in flux and generated e.m.f. This is used in electric arc welding sets where approximately 110 V is required to strike the arc but only 20 V to maintain the arc whilst welding.

ARMATURE REACTION

Further work on commutation

Figure 6.45(a) shows one coil of the lap winding originally shown in *Figure 6.19*. The negative brush is resting on segment 4 of the commutator. As viewed the current direction in the coil 3, 3′ is anti-clockwise. In *Figure 6.45(b)* the conductors have moved a distance equivalent to one half of a commutator segment to the right. Both coil sides are now between poles and there is no induced e.m.f. in the coil. The brush now shorts out the coil since it spans segments 3 and 4.

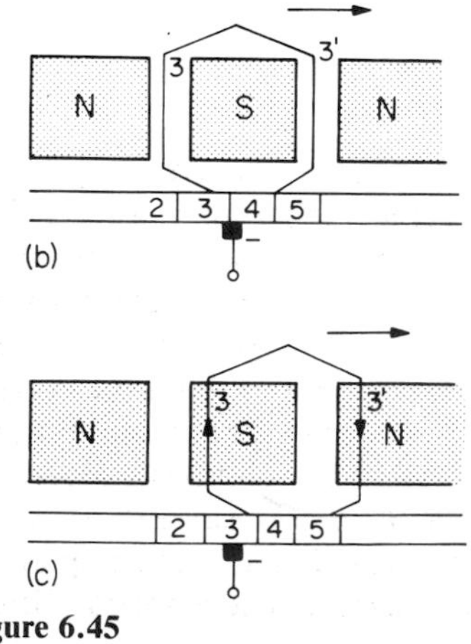

In *Figure 6.45(c)* the coil has again moved to the right a further half segment of the commutator and both coil sides are under poles. The currents are now as shown, the direction being clockwise as viewed.

This means that in the time taken for the armature to move from position (a) to position (c) the current in the coil must reverse in direction. If the coil has a large inductance which is likely since it is embedded in a large mass of iron, and again if the current is large, there will be a considerable induced voltage in the coil during the period of reversal. This is known as reactance voltage.

The reactance voltage

$$e = L\frac{\mathrm{d}i}{\mathrm{d}t} \text{ volts}$$

where L = inductance of the circuit in henrys and $\mathrm{d}i/\mathrm{d}t$ = rate of change of current in amperes per second.

Severe sparking will occur at the surface of the commutator if the current reversal, or commutation, is not complete by the time that the commutator segment moves from under the brush.

During commutation the sides of the coil must be outside the magnetic field so that the generated e.m.f. is zero and to achieve this

Coil motion

(a)

(b)

(c)

Figure 6.45

the brushes are moved round the commutator until they are located on what is termed the neutral axis.

Armature reaction

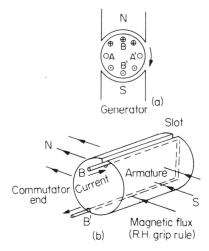

Figure 6.46

In *Figure 6.46(a)* a simple armature is shown revolving between two magnetic poles. The current directions are shown. The coil A,A' which lies horizontally has no e.m.f. induced in it and the current in it is undergoing commutation. The coil B,B' which is in a vertical plane at this instant is shown separately in *Figure 6.46(b)*. The current in conductor B flows away from the commutator end along the length of the armature, crosses over at the back and then flows back in conductor B'. This current produces a magnetic north pole on the left hand side of the armature. The other two current carrying coils will add to the effect. The strength of this cross-magnetic field will depend on the number of turns and the value of armature current.

In *Figure 6.47*, two ways of illustrating the effect of the armature or cross flux are shown. In *Figure 6.47(a)* the distorting effect of the current in the two sides of the coil B,B' is shown. In *Figure 6.47(b)* the vector addition of the armature and main fluxes can be seen. The resultant flux has been displaced by $\theta°$ from the vertical. The effect of the armature flux on the main flux is called armature reaction. This term must not be confused with reactance voltage which is quite different.

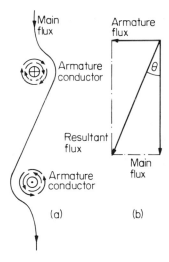

Figure 6.47

In a multi-pole machine where the space between the poles is much smaller than in the two-pole machine this twisting of the field can bring into the field conductors which were out of it and in which commutation is taking place. Severe sparking results. The position for the brushes to be situated in order to obtain satisfactory commutation has now moved $\theta°$ in a clockwise direction from the former position. The new position is known as the magnetic neutral axis and its position is a function of the armature current since the amount of field distortion is dependent on this current. In some older machines brush moving gear was provided so that an operator could adjust the position of the brushes as the load changed. This is not always convenient and other methods are generally employed to achieve satisfactory commutation under all load conditions.

Compensating windings and interpoles

One method of overcoming the effect of armature reaction is to fit compensating windings to the machine. These carry the armature current and so are series connected. They are arranged to create magnetic poles which oppose those set up by the armature currents so that at any instant the resultant cross field is zero. This type of winding is very expensive and is therefore only used on some special machines.

A more usual alternative method is the use of interpoles. These are very thin poles carrying armature current and these are situated between the main poles. They are wound to give a polarity which anticipates the action of the next main pole. They assist the reversal of the current in the coil undergoing commutation. The interpole not only neutralises the effect of armature reaction local to the conductors undergoing commutation but overcomes the effects of reactance voltage.

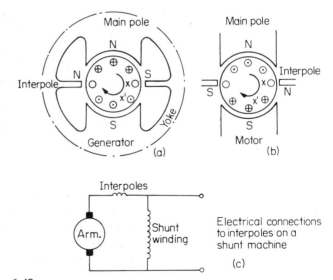

Figure 6.48

When considering the action of the interpoles it must be remembered that they always act in the generating mode whether they are fitted to motors or generators. Fleming's right hand rule always applies to them. They assist in inducing a voltage of the correct polarity before the conductor actually moves under the next main pole. In *Figure 6.48(a)* the simple generator rotating in a clockwise direction has the armature current directions as shown. As the conductor on the right moves from a position X to X' the current must be established out of the paper as viewed. In order to assist this to occur the interpole field must be as shown. *Figure 6.48(b)* shows the interpole polarities for a motor with the same rotation.

Changes in characteristics due to armature reaction

Armature reaction causes the magnetic field in a d.c. machine to be twisted round which concentrates the flux into one side of each pole piece so that this side may be driven into saturation as a result. Since the top limit to flux density in iron is in the region of 1.6 to 1.8 tesla this may result in a reduction in total flux.

Example 10. The normal flux density of the poles in a d.c. generator on no load is 1.1 T. Each pole has a cross-sectional area of 0.05 m². The effect of armature reaction at a particular load is to concentrate all the flux into one half of the pole face area. The saturation flux density is 1.8 T. Calculate the new value of flux per pole and the per unit reduction in flux at this load.

Total flux per pole = flux density × area of the pole face.
$$= 1.1 \times 0.05$$
$$= 0.055 \text{ Wb.}$$

Concentrating all this flux into one half of the pole face could be expected to give a flux density of

$$\frac{0.055}{0.05/2} = 2.2 \text{ T.}$$

This is above saturation. The flux density cannot rise above 1.8 T. At 1.8 T and an area of 0.025 m² the flux per pole $= 1.8 \times 0.025 = 0.045$ Wb. This is a reduction of $0.055 - 0.045 = 0.01$ Wb.

$$= \frac{0.01}{0.055} = 0.19 \text{ p.u. reduction.}$$

In practice the reduction is seldom if ever this large, but the example serves to demonstrate the principle.

This reduction in flux will cause a further reduction in the output voltage of d.c. generators in addition to those already discussed.

In the case of motors there will be a speed increase as the flux is reduced and in addition, if the load torque remains constant, the armature current will increase. These effects can be explained by a consideration of the e.m.f. and torque equations.

LOSSES IN DC MACHINES

(See *Figures 6.27, 6.29* and *6.42* for connections and symbols.)

Field loss

In the shunt machine the field loss $= VI_f$ or $I_f^2 R_f$ watts.
In the series machine, since the field winding carries the armature current, field loss $= I_a^2 R_{sf}$ watts.
The field loss is a function of the number of ampere turns necessary to magnetise the pole pieces and so is dependent on the grade of steel used.

Armature loss

(a) $I_a^2 R_a$ watts loss in the coils of the armature. For a particular load the armature current is constant so that this loss is directly proportional to the resistance of the winding.

(b) Iron losses. Since the armature is subjected to alternating magnetisation as it passes under north and south poles, there will be hysteresis and eddy current losses in the steel. The hysteresis loss is a function of: (1) the maximum flux density (2) the grade of steel used and (3) the speed of rotation of the armature. Eddy current losses are a function of: (1) the maximum flux density (2) the thickness of the laminations used (3) the resistivity of the steel and (4) the speed of rotation.

Commutator losses

(a) Resistance loss. There is resistance between the brush surface and the commutator and the loss depends on the grade and quality of the brushes used. The volt drop tends to be constant at about 1 V per brush so that this loss is approximately (2 volts × total armature current) watts, irrespective of the type of armature winding. (b) Friction loss. There is rubbing friction between the brushes and the commutator and the power loss depends on the coefficient of friction and the rubbing speed. The rubbing speed is a function of the diameter of the commutator and the speed of rotation.

Bearing friction

The armature is supported in bearings and these may be of the ball, roller or sleeve types according to motor application. The frictional loss is a function of the speed of rotation and the type of bearing.

Windage

There will be friction between the surface of the armature and the air in the casing. In addition power will usually be required to drive the cooling fan. The power loss involved depends on the type and size of fan and on the running speed.

THE POWER FLOW DIAGRAM

For a motor

Note: The term 'Armature circuit' includes any series field windings and interpoles present.

Starting from the top of *Figure 6.49* we have the total input to the motor. From this is subtracted the shunt field loss if applicable. VI_a watts is supplied to the armature circuit where the losses are I_a^2 (total resistance of the armature circuit) watts plus the brush resistance loss. The power to create the gross torque is EI_a watts. The gross torque drives the armature against brush and bearing friction, windage and the retarding torque due to hysteresis and eddy currents in the armature. After all the losses have been provided we have the useful output which drives the connected mechanical load.

For a generator

Starting at the bottom of *Figure 6.49* we have the total input which is being provided by an engine of some form. The input must provide all the mechanical losses leaving EI_a watts from which the armature circuit losses must be subtracted. This leaves VI_a watts as the output from the armature circuit. Finally in the shunt machine its own field loss must be deducted leaving the useful power output at the top.

In both the motor and the generator case the input is greater than the output and the ratio output/input is the efficiency of the machine.

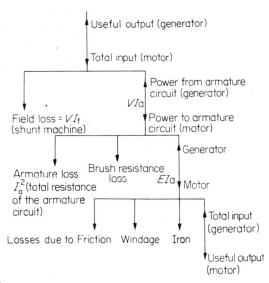

Figure 6.49

Typically this will lie between 0.6 and 0.85 according to the rating of the machine.

PROBLEMS FOR SECTION 6

11 What is the function of a commutator in a d.c. machine?
12 What is the basic difference between a lap and a wave winding on a d.c. armature?
13 Which type of winding would you choose for: (a) high current, low voltage application (b) high voltage, low current application.
14 Calculate the value of the generated e.m.f. delivered by the following six-pole d.c. generators.
   (a) Lap wound: $\Phi = 20\,\text{mWb/pole}$,
       number of slots on the armature = 86
       conductors per slot = 5, speed = 15 rev/s
   (b) Wave wound: $\Phi = 25\,\text{mWb}$, number of slots = 24,
       conductors per slot = 10, speed = 20 rev/s.
15 A d.c. machine has an armature resistance of $0.5\,\Omega$. It is connected to 500 V mains. When drawing 20 A from the supply it runs at 25 rev/s. At what speed must it run as a generator in order to deliver 20 A to the electrical system? The flux remains constant.
16 The armature of a six-pole, lap wound d.c. motor has 54 slots and 8 conductors in each slot. The total flux per pole is 0.05 Wb. The resistance of the armature is $0.3\,\Omega$. When connected to a 240 V supply, at a particular load the armature current is 20 A. Calculate: (a) the speed under these conditions (b) the value of flux per pole required to increase the speed to 15 rev/s all other conditions remaining unchanged.
17 Draw simple circuit diagrams illustrating (a) the series connection (b) the shunt connection and (c) a compound connection.

18 Sketch speed/armature current and torque/armature current characteristics for motors (a) and (b) in Q 17 above.

19 A d.c. shunt motor has an armature resistance of 0.3 Ω. It is to be started using a resistance starter. The supply voltage is 250 V. Determine the value of starting resistance for the armature circuit so that the current at standstill does not exceed 25 A.

If at a particular speed the back e.m.f. has risen to 125 V, to what value should the starting resistance be reduced to give the original value of current?

20 What relationship must exist between armature coils and field poles in order to achieve good commutation? What design features are incorporated in many machines to assist commutation?

21 A d.c. machine has the following parameters:
$R_a = 0.5$ Ω $R_f = 220$ Ω. Brush resistance loss = 100 W, Friction, windage and iron losses = 1500 W. Supply voltage = 440 V.
Determine the output power and efficiency of the machine when operating as a motor and the armature current is 50 A.

22 For the machine in Question 21 operating as a generator supplying 50 A to the electrical system, determine the total power input and efficiency.

23 What are the advantages of starting a d.c. motor using electronic means rather than by the inclusion of rotor resistance, especially when the power supply is a large traction battery?

24 What are the advantages of driving a d.c. motor from a.c. mains through a controlled rectifier as compared with driving the same motor from d.c. mains?

7 Illumination

Aims: At the end of this chapter you should be able to:

State that light is emitted by hot tungsten wire in an incandescent
   lamp and an electrical discharge in a gas.
Compare efficiencies and colour spectra of various lamps.
Define flux, intensity, illuminance and luminance.
State the inverse square law.
Calculate the illuminance of a surface using $E = I \cos \theta / d^2$ *and a*
   polar diagram.
Calculate the number of luminaires needed to provide uniform
   lighting using the lumen method together with maintenance and
   utilisation factors.
Describe the effects of diffusing and prismatic luminaires.
State the conditions which affect the amount of daylight reaching a
   point in a room.

White light may be resolved into a band of colours called the spectrum ranging through red, yellow, green, blue and violet by passing it through a glass prism as shown in *Figure 7.1*. The same effect occurs naturally when sunlight falls on water droplets, forming a rainbow against a background of dark cloud.

These colours are due to electromagnetic radiations similar to radio waves but of a much higher frequency.

Red light has a frequency of 4×10^{14} Hz

Violet light has a frequency of 7.5×10^{14} Hz

By comparison, radio transmissions have frequencies starting at about 1×10^5 Hz in the long-wave band extending to about 1×10^8 Hz in the very-high frequency band. Television signals need a higher frequency still, for example 8×10^8 Hz, and in some ways these act similarly to light, it being difficult to receive signals over a hill for example.

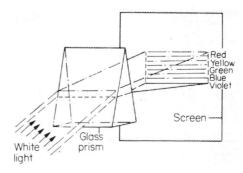

Figure 7.1

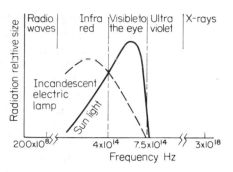

Figure 7.2

The speed of all electromagnetic radiations such as light and radio is 3×10^8 m/s (approximately).

At frequencies slightly lower than red there is infra-red radiation which, although invisible, can be detected by the human skin since it is warming. At the other end of the spectrum at frequencies above that of violet light, there is ultra-violet radiation which also has an effect on the human skin, turning it brown as a means of protection. Excessive doses of ultra-violet radiation can be dangerous, burning the skin and damaging the eyes.

The essential source of light and heat for the earth is the sun which provides radiations from ultra-violet, right through the visible spectrum, and into the infra-red region. Much harmful radiation is absorbed by the earth's atmosphere and never reaches the ground, but in areas where this is very clear there are still dangers and eye protection is worn by mountain climbers and skiers for example.

Many solid bodies emit light when raised to a sufficiently high temperature. The ordinary incandescent light bulb, which contains a fine tungsten wire being heated by the passage of an electric current, is a good example. The energy supplied is converted into heat and light and *Figure 7.2* shows the various frequencies of radiations in the output from such a bulb. Note the relatively large amount of heat produced and how little of the radiation is in the visible region.

LIGHTING STANDARDS

Since measurements of illumination were first made it has been necessary to have a standard with which the various light sources could be compared.

The original standard was the wax candle. The exact level of illumination was difficult to reproduce each time it was required since the grade of wax, the material for the wick and its burning length had to be specified and accurately controlled. This standard was followed by a lamp burning pentane gas and later by a standard incandescent lamp.

In 1948, with the introduction of SI units, the platinum standard was adopted. It has already been noted that when solid bodies are heated they can emit light. The nature of the light emitted depends on the temperature and this is rather difficult to determine accurately. The problem is overcome by using a material which emits light at a temperature corresponding to that which it solidifies. It therefore exploits the phenomenon that when materials change state, from liquid to solid or liquid to gas, they do so at constant temperature. A common example is that of water as it freezes into ice or evaporates into steam.

Platinum melts or freezes at 1773°C and at this temperature emits light since it is white hot. The platinum, contained in a small tube within an insulated box, is melted electrically. During the time it is solidifying the temperature remains constant so that the light intensity at a small viewing hole at the top of the box is constant. This light is used as the standard and other sources of light are compared with it. Another reason that platinum is ideal for the standard is that it remains chemically unchanged during successive melts.

LUMINANCE AND LUMINOUS INTENSITY

The luminance of the viewing hole in the standard is defined as 60 candelas per square centimetre (60 cd/cm²). Let us examine what this means.

Luminance can be interpreted as a measure of the discomfort caused by looking at a light source. Imagine trying to look directly at the noonday sun. It will cause pain and may indeed damage the eye if there is prolonged exposure. The sun has very high luminance, estimated to be about 160 000 cd/cm² at noon in summer at Greenwich. In contrast a 1500 mm, 80 W fluorescent tube commonly used for lighting offices and shops has a luminance of about 1 cd/cm² and this may be viewed without risk. Imagine now being in a completely darkened room when a small hole is cut in one of the blinds through which the bright sun may be viewed. The amount of light coming into the room through that hole would be minimal and it would be difficult to see things in the room. If however a single fluorescent tube is switched on in the room, far more light is provided than by the sun shining through the small hole.

It can therefore be deduced that light radiation capability or luminous intensity is the product of luminance and area. Thus in the above example the sun with its very high luminance is projecting light through a small area and thereby providing less light than a fluorescent tube of low luminance but with a large surface area.

Since the platinum standard has a luminance defined as 60 cd/cm², a hole in the top of the standard having an area of 1/60 cm² would result in a light radiating capability of 1 cd. The symbol for light radiating capability or luminous intensity is I so in this case, $I = 1$ cd.

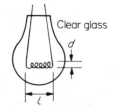

Clear glass

Projected filament area $l \times d$

Figure 7.3

Example 1. The filament of a 100 W clear incandescent lamp has a luminance of 650 cd/cm². If the filament has a projected area of 0.1 cm² as shown in *Figure 7.3*, what is the light radiating capability in candelas?

I = luminance × area
   $= 650 \times 0.1$
   $= 65$ cd.

Instead of using clear glass, the interior surface may be treated to create the pearl lamp. This makes the filament appear to have a larger area. If the light radiating capability is unchanged the luminance must have decreased. Where the lamp can be viewed directly, the pearl lamp would be preferred. Alternatively where the lamp is enclosed in some form of shade or diffuser, making direct viewing impossible, the clear lamp would be quite suitable.

THE LUMEN AND THE STERADIAN

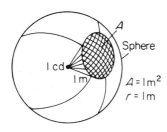

Figure 7.4

Figure 7.4 shows a light source with uniform luminous intensity of 1 candela at the centre of a sphere of radius 1 metre. A uniform source is considered to radiate light equally in all directions but this is not achievable absolutely in practice since all lamps have a dark area where the energy is fed in.

If we consider an area A on the surface of the sphere of 1 square metre, the solid angle so formed is called a steradian, symbol ω.

The amount of light from the source which passes through the cone, i.e. within the solid angle of 1 steradian, is 1 lumen.

The amount of light is called luminous flux and has symbol Φ. Since the radius of the sphere is 1 metre, its surface area can be calculated using

Area $= 4\pi r^2$ metre^2,

Area of the sphere $= 4\pi$ metre^2 and each square metre forms 1 sheradian within the sphere.

Therefore the total solid angle within the sphere $= 4\pi$ steradians.

Because there is a luminous flux of 1 lumen in each steradian from a 1 candela source, the total light flux from such a source must be 4π lumens. In *Figure 7.4*, 1 lumen is falling on 1 square metre of area giving a level of illumination or illuminance at the surface of 1 lux. Illuminance has the symbol E. Hence $E = 1$ lux.

In general where the source has luminous intensity I candela, the total luminous flux through ω steradians is $I\omega$ lumens.

Example 2. A lamp has a uniform luminous intensity of 100 cd. What is the total light output of the lamp in lumens?

If all of this light is reflected on to an area of 10 m^2, what is the illuminance of that area?

Total luminous flux $= 100 \times 4\pi$ lumens
$= 1256.7\,\text{lm}$

On an area of 10 m^2 this results in an illuminance of

$$\frac{1256.7}{10}\ \text{lm m}^2$$

Since the number of lumens falling on 1 m^2 has been defined as the illuminance in lux,

$E = 125.7\,\text{lx}.$

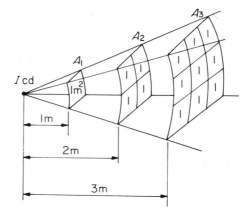

Figure 7.5

THE INVERSE SQUARE LAW

In *Figure 7.5*, the source of illumination has a luminous intensity of *I* candela. The level of illumination at surface A_1 is due to *I* lumens on 1 square metre since the solid angle is 1 steradian.

$E = I$ lux

The light rays are all meant to be at right angles to the surface being illuminated. If the surface A_1 is removed so that all the light rays which illuminated it now travel two metres to illuminate surface A_2, the luminous flux has spread over four square metres. There are now *I* lumens on 4 square metres.

$$E = \frac{I}{4} \text{ lumens per metre}^2 \text{ or lux.}$$

For surface A_3 the same luminous flux is distributed over nine square metres.

$$E = \frac{I}{9} \text{ lux.}$$

Generally then, the illuminance $E = I/d^2$ lux where *d* is the distance of the illuminated surface from the source in metres.

THE COSINE LAW

In the previous section the light rays were all normal to the surface being illuminated. In the case of street lighting for example this would be true only for the position immediately below each lamp.

Consider a single lamp and a point on the road surface some distance from the base of the lamp as shown in *Figure 7.6*. At this range the rays of light are considered to be very nearly parallel. The level of illumination on the area of one square metre shown normal to the light rays is I/d^2 lux. As this area is tilted through the angle $\theta°$ so that it lies in the horizontal plane, the same luminous flux now illuminates an area of $1/\cos \theta$ metre^2. The level of illumination on this increased area now becomes

$$\frac{I}{d_2} \div \frac{1}{\cos \theta} = \frac{I \cos \theta}{d^2} \text{ lux.}$$

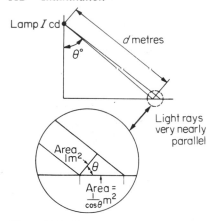

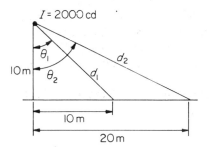

Figure 7.6

Figure 7.7

Example 3.   Calculate the level of illumination (illuminance) on a horizontal surface due to a single lamp with uniform light radiating capability of 2000 cd mounted 10 m above the surface at points:

(a) Immediately below the lamp.
(b) 10 m from this point on the horizontal surface.
(c) 20 m from the original point on the horizontal surface.

(a) Since the lamp has uniform light radiating capability, the value 2000 cd may be used throughout.
     Immediately beneath the lamp the light rays fall normally on the surface.

$d = 10$ m

$$E = \frac{2000}{10^2} = 20 \text{ lux}$$

(b) $\tan \theta_1 = \frac{10}{10} = 1$. Thus $\theta_1 = 45°$. This may be obtained by scale drawing if required.

By Pythagoras' theorem: $d_1^2 = 10^2 + 10^2 = 200$

$$E = \frac{2000}{200} \cos 45°$$

$$= 10 \times 0.707$$
$$= 7.07 \text{ lux.}$$

(c) $\tan \theta_2 = \frac{20}{10} = 2$ Thus $\theta_2 = 63.5°$
     $d_2^2 = 10^2 + 20^2 = 500$

$$E = \frac{2000}{500} \cos 63.5° = 1.784 \text{ lux.}$$

Example 4.   A light fitting with uniform light radiating capability is mounted 12.5 m above level ground. At a point on the ground 20 m from the base of the lamp standard the illuminance is 2 lx. Determine: (a) the light radiating capability of the lamp in candelas, (b) the total lumen output of the lamp.

THE POLAR DIAGRAM

A lamp without any means of directing the light produced, radiates a good deal of its light in directions other than those required, on to the ceiling for example, or in the case of road lighting, up into the sky. If a reflector or some form of shade or lantern is used in conjunction with the lamp, light can be radiated efficiently in those directions required.

The light radiation capability of the fitting is now no longer uniform and in order to calculate the illuminance of a surface it will be necessary to use the value of luminous intensity in the direction towards the surface, e.g. along the lines inclined at θ_1 and θ_2 to the vertical in *Figure 7.7*. A polar diagram provides information on the luminous intensities in particular directions.

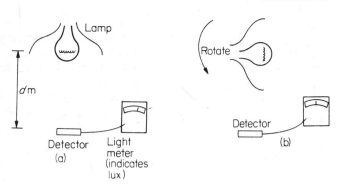

Figure 7.8

Polar diagrams for particular lamps and reflectors may be obtained from the lamp manufacturers. Alternatively they can be obtained experimentally by using the following procedure.

A lamp, optionally with reflector, is mounted in a room with matt black walls and in which there is no other source of illumination. A light meter is situated at a known distance from the lamp filament. This distance is often one metre to make the calculations easy. The arrangement is shown in *Figure 7.8*.

The light meter measures illuminance due to the lamp at the light-sensitive surface of the detector.

$$\text{Illuminance } E = \frac{I}{d^2} \text{ lux}$$

If the distance involved is one metre then the meter reading in lux is numerically equal to the luminous intensity I of the source. In *Figure 7.8(a)* the luminous intensity in a direction along the axis of the lamp is being found.

If the lamp is turned through 90° as shown in *Figure 7.8(b)* the luminous intensity of the lamp at right angles to its axis is being found. The process is carried out at intervals of 10° or 20° throughout a complete revolution. Thus the luminous intensity of the lamp and its reflector in any particular direction is evaluated. A plot of

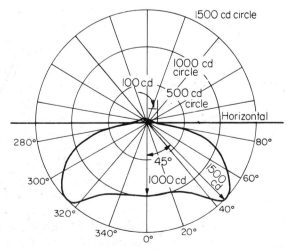

Figure 7.9

the results on polar graph paper is referred to as the polar diagram. Such a curve for the lamp and reflector in *Figure 7.8* is shown in *Figure 7.9*

The light radiating capability of the fitting is scaled from the origin 0 in the direction required. For instance, directly downwards it is 1000 candelas whilst in the horizontal direction it is only 100 cd. At 45° to the horizontal it is 1500 cd.

The light radiating capability above the horizontal is virtually zero since this particular fitting is designed to project most of the light produced on to the floor.

REFLECTION AND REFRACTION OF LIGHT

Light may be directed to the required area by reflection or refraction. Reflection of light takes place to some extent at most surfaces. If the surface is highly polished, as with a mirror, the reflection is said to be 'specular'. The angle of incidence is equal to the angle of reflection as shown in *Figure 7.10*. Specular reflection is used in the parabolic reflector which can be found in floodlighting fittings or luminaires, in cinema projectors and searchlights etc.

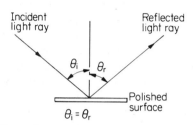

Figure 7.10

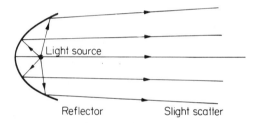

Figure 7.11

The white enamelled steel industrial reflector reflects most of the light downwards as shown in *Figure 7.12*, distributing it over a wide area. Where the surface is matt finished, diffused reflection takes place. This gives a softer lighting effect although it it is not so efficient as specular reflection due to light absorption by the surface. The colour of the surface also affects reflection, light colours obviously reflecting more light than dark ones.

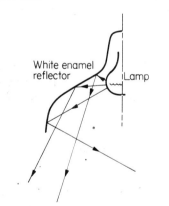

Figure 7.12

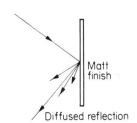

Figure 7.13

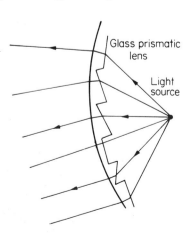

Figure 7.14

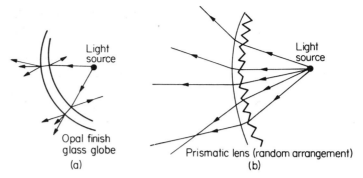

Figure 7.15

Refraction or bending of light rays takes place when they pass through a prism as shown in *Figure 7.1* The reflecting lantern or luminaire is used for street lighting and in motor vehicle headlamps. An array of prisms may be seen on the inside of the glass as shown in cross-section in *Figure 7.14*. By a combination of these methods a polar diagram of almost any shape can be produced enabling light to be projected in the direction where it is most needed. If the object of a luminaire is to diffuse the light so as to reduce the luminance of the source, then diffusing glass globes may be used. These will be finished pearl or opal on the inside or again may make use of much smaller prisms which have the effect of scattering the light as shown in *Figure 7.15(b)*.

USE OF THE POLAR DIAGRAM

Suppose the surface level of illumination or illuminance provided by a lamp in a specific type of luminaire is required. The polar curve is plotted or obtained from the manufacturer. A typical example is shown in *Figure 7.16*.

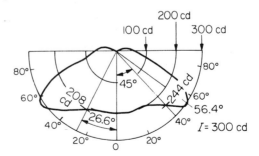

Figure 7.16

The following example illustrates the method to be adopted.

Example 5. A single lamp in a reflector has a polar diagram as shown in *Figure 7.16*. It is mounted on a lamp standard at a height of 10 m above a horizontal flat plane. Calculate the illuminance on the plane:
(a) Immediately below the lamp,
(b) 10 m from the base of the lamp standard.

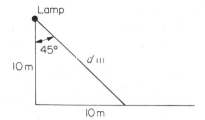

Figure 7.17

The first step is to draw the arrangement of lamp, standard, and surface to be illuminated as shown in *Figure 7.17*. The lamp is considered to be situated at the origin of the polar diagram. The light illuminating the spot immediately below the lamp is shining directly downwards through zero degrees on the polar diagram. By direct reading from the curve, the luminous intensity in this direction is 200 cd.

(a) $E = \dfrac{I}{d^2} \cos \theta$ lux and in this case $\theta = 0°$.

$$E = \frac{200}{10^2} = 2 \text{ lux.}$$

(b) At a distance of 10 m from the base of the standard, the light is projected at an angle of 45° from the vertical

$$d^2 = 10^2 + 10^2 = 200.$$

From the polar diagram, at 45° to the vertical, the luminous intensity of the source is 244 cd.

$$E = \frac{244}{200} \cos 45° \text{ according to the cosine rule.}$$

$$E = 0.863 \text{ lux.}$$

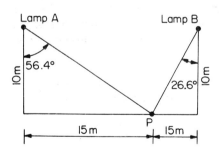

Figure 7.18

Example 6. Consider the case of two lamps 20 m apart as shown in *Figure 7.18*

The mounting heights and polar diagram for each lamp are as in the previous example. Calculate the level of illumination on a line joining the bases of the two standards at a distance of 15 m from one of them. This is point P in *Figure 7.18*.

The illumination at point P comes from two sources. Consider them separately.

From lamp A: $E = \dfrac{300}{10^2 + 15^2} \cos 56.4°$

$$= 0.51 \text{ lux.}$$

From lamp B: $E = \dfrac{208}{10^2 + 5^2} \cos 26.6°$

$$= 1.48 \text{ lux.}$$

Total illuminance due to both sources = 1.48 + 0.51
$$= 1.99 \text{ lux.}$$

Example 7. Calculate the illuminance due to both lamps directly under one of the lamps in Example 6. The polar diagram for each lamp is again as in *Figure 7.16*

INDIRECT ILLUMINATION OF A SURFACE

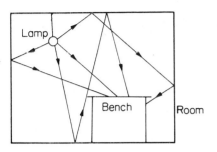

Figure 7.19

The previous examples using the polar diagram show how the illumination at a point may be calculated when that illumination is due to a single source, with no reflecting surfaces nearby, e.g. as in road lighting. If there are two or more sources, again with no reflecting surfaces nearby, then the illuminance present is the sum of the values due to individual sources. With indoor illumination however, this method cannot be used since the distances are small and much of the illumination is the result of reflection from ceilings and walls.

Consider a workshop which requires a given illuminance at the bench tops. The benches receive light in any number of ways as shown in *Figure 7.19*. Some comes directly from the fitting and some by reflection from the walls, ceiling and floor. Note that light can be reflected several times, losing a little by absorption on each occasion.

The amount of light received after reflection depends on the colour and cleanliness of the surfaces involved. It also varies according to the height of the ceiling and the shape of the room. The arrangement of the reflecting surfaces, their condition and colour is taken into account by the use of a 'coefficient of utilisation'.

The coefficient of utilisation is defined as

$$\frac{\text{Luminous flux arriving at the working surface}}{\text{Total luminous flux supplied by the lighting fittings}}$$

This has been determined experimentally by the Illuminating Engineering Society for many different combinations of colour and room proportions and may be found from reference tables in the IES code. The IES also specify the requirement for illuminance at the working surface for various types of work and the following table quotes some of these.

Workshops and factories	Illumination recommended in lux
Average for general work	100–150
Assembly – large work	100
Radio assembly	200
Engraving	500
Warehouses	50
Drawing offices	300

Example 8. A workshop 30 m × 40 m is to be illuminated using gas-filled tungsten filament lamps in standard dispersive reflectors. They are situated 2.5 m above the working plane on which an illuminance of 100 lux is required.
The coefficient of utilisation is 0.6.
Calculate the total luminous flux required from the fittings.

100 lux = 100 lumens on each square metre.
Area to be illuminated = 30 × 40 m²
$$= 1200 \text{ m}^2$$

Luminous flux at the working surface $= 1200 \times 100$
$$= 120\,000\,\text{lm}.$$

Since the coefficient of utilisation

$$= \frac{\text{Luminous flux at the working surface}}{\text{Total input of luminous flux}}$$

$$0.6 = \frac{120\,000}{\text{Input}}$$

Hence, input $= \dfrac{120\,000}{0.6}$

$$= 200\,000\,\text{lm}.$$

During the life of an electric lamp its light output decreases due to ageing of the filament or of the phosphors employed in fluorescent tubes. Lamp reflectors also become less efficient due to an accumulation of dust. Inevitably some lamps fail altogether. In a factory it is not usually economic to replace single lamps especially if they are rather inaccessible. It is better to wait until either 5 per cent of the installed lamps have gone out or all the lamps have been in operation for their design life before replacing all the lamps at the same time. This can usually be done when the factory is not in production.

Where a minimum level of illumination is specified, allowance is made for deterioration of lamps by providing a greater level of illumination than this minimum when the installation is new. At the end of the design life of the lamps the output will still be maintained at or above the minimum requirement.

A depreciation factor is used to calculate the required original input. If the illuminance in the workshop of the previous Example 8 must never fall below 100 lux and a depreciation factor of 1.3 is recommended, then the input when new must be $1.3 \times 200\,000$ lumens.

Alternatively a maintenance factor may be used.

$$\text{Maintenance factor} = \frac{\text{light output at the end of the design life}}{\text{light output when new}}$$

Again using data from the previous example, the maintenance factor would have been 0.77.

$$0.77 = \frac{200\,000}{\text{light output when new}}$$

Light output when new $= \dfrac{200\,000}{0.77} = 260\,000$ lumens
$(= 1.3 \times 200\,000)$

$$\text{Depreciation factor} = \frac{1}{\text{Maintenance factor}}$$

Clearly, whenever the maintenance or depreciation factor is used in calculations it must give an original light input which exceeds that at the end of the design life.

It is now necessary to examine the numbers and arranagement of lamps by which the required illuminance may be provided.

Data for gas filled tungsten filament lamps is as follows:

Power, watts	60	100	200	300	500
Total output, lumens	576	1135	2600	4140	7500

Since a 60 w lamp gives 576 lumens, to provide 260 000 lumens would require 260 000/576 = 452 lamps.

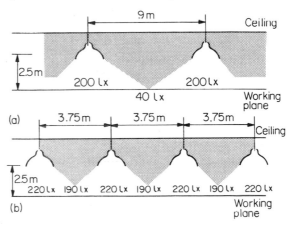

Figure 7.20

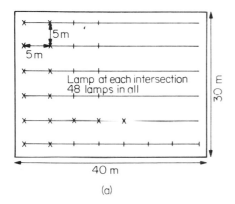

(a)

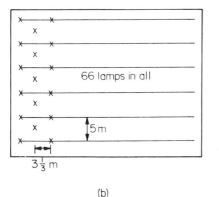

(b)

Figure 7.21

Alternatively, using 500 W lamps, 260 000/7500 = 35 lamps would be required.

The effect of having a few large lamps widely spaced is to leave areas of low illuminance between the lamps. This may be seen in *Figure 7.20(a)*. Generally it is found that spacing the lamps between 1.5 and 2 times their mounting height above the working plane produces an acceptable degree of variation. This is shown in *Figure 7.20(b)*. Using lamps closer together than this increases the installation costs. Two suitable arrangements of lamps are shown in *Figure 7.21*.

The arrangement in *Figure 7.21(a)* uses 48 lamps each with 500 W rating. The total electrical loading = 48 × 500 = 24 000 W and the luminous flux provided = 48 × 7500 = 360 000 lumens.

The arrangement in *Figure 7.21(b)* uses 66 lamps each with 300 W rating. The total electrical loading = 66 × 300 = 19 800 W whilst the total luminous flux = 66 × 4140 = 273 200 lumens.

Note that it is not generally possible to satisfy the requirements exactly without having an asymmetrical arrangement of lamps and this would have an unsatisfactory appearance.

TYPES OF LAMP

The incandescent lamp

These lamps have a tungsten filament heated within a glass bulb which has been either exhausted to near-perfect vacuum or contains a small amount of inert gas. It radiates energy over a continuous spectrum as shown in *Figure 7.2*. The proportion radiated as light

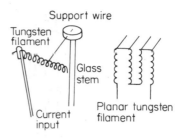

Figure 7.22

increases with filament temperature but at best this is only a small fraction of the available energy.

Figure 7.22 shows two possible filament arrangements, one for general service lamps and the other for film projectors. Outputs vary from about 9 lumens per watt for the 40 W size to about 14 lumens per watt for the 500 W size.

The high-pressure mercury discharge lamp

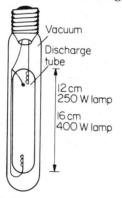

Figure 7.23

In this lamp, current is caused to flow through a small tube in which there is little inert gas and a few droplets of metallic mercury. The current flows at first in the gas producing enough heat to vaporise the mercury. The ionisation of the mercury vapour produces light. The pressure when hot is about 1 bar.

The space between the inner discharge tube and the outer bulb is partially evacuated to minimise the heat loss from the lamp. This is necessary to keep the mercury in vapour form.

The colour of the light is blue/green with no additional radiation corresponding to red and yellow present. However, there is ultra-violet present and if the inside of the outer bulb is coated with a suitable phosphor, the ultra-violet emission causes this to emit a red component so making the light emitted more nearly equivalent to daylight. The uncompensated spectrum is shown in *Figure 7.26*. The light output is about 40 lumens per watt.

The low-pressure mercury (fluorescent) tube

This lamp also contains mercury but runs at a much lower pressure than the type just described. It is also larger physically, being typically 1500 mm long at 65 or 80 W rating. It runs much cooler at around 40°C. The discharge from the mercury under these conditions contains a substantial amount of ultra-violet. The inside of the tube is coated with phosphors which emit visible radiations (light) as a result of being excited by the ultra-violet radiations from the gas. The colour can be altered by the selection of phosphors, a technique employed in the manufacture of colour television tubes. The light output of the tube varies between 40 and 60 lumens per watt according to colour, 'blue' (north light) tubes giving less light than 'warm white' tubes for the same rating.

The low-pressure sodium discharge lamp

This lamp consists of a U-tube containing a little sodium metal and neon gas at very low pressure. When the supply is switched on, the neon gas conducts current and a bright red light is emitted. Slowly the sodium boils off into the near vacuum of the tube and the ionised sodium vapour emits bright yellow light. The spectrum is shown in *Figure 7.26*. As with the high-pressure mercury lamp, heat must be conserved and therefore the U-tube is supported in a second

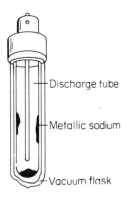

Discharge tube

Metallic sodium

Vacuum flask

Figure 7.24 Low pressure sodium lamp

vacuum flask. If this fractures, the light reverts to the red colour as the sodium solidifies. The light output is about 130 lumens per watt. This lamp is used for road lighting where the colour of the light is not too significant.

The halogen lamp

The conventional tungsten filament decreases in cross-sectional area by evaporation. The tungsten condenses on the bulb wall, so reducing the light output. The glass bulb is made large and generally the lamp burns with its cap at the top. Convection currents inside the lamp carry the gaseous tungsten up into the neck of the lamp where it condenses out. The loss of light output due to the blackening of the glass is thus limited.

The halogens are chlorine, bromine, iodine and fluorine. When a halogen is added to the gas inside the lamp and if certain temperature conditions are satisfied, a regenerative chemical reaction may be set up. The tungsten evaporates from the filament as before but, as

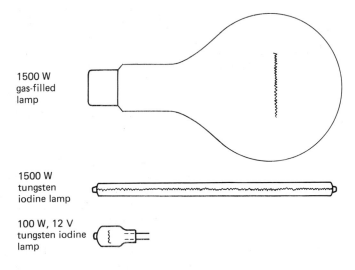

1500 W gas-filled lamp

1500 W tungsten iodine lamp

100 W, 12 V tungsten iodine lamp

25 cm

Figure 7.25

it approaches the bulb wall and its temperature falls somewhat, the tungsten combines with the halide to form a compound, tungsten halide. This compound does not condense on the glass but circulates within the bulb until it comes into contact with the hot lamp filament once more, when the halide is released and the tungsten deposited on the filament. Iodine is the halide currently used, which requires that the filament runs at about 1800°C and the bulb walls are maintained at 250°C. To ensure that the bulb walls attain this temperature, the lamp must be made very much smaller than the conventional type, but this is no longer significant since there is no blackening of the glass. Because of their reduced size, these lamps can be placed at the centre of focus of a parabolic reflector and they are found in motor vehicle headlamps and floodlights for public areas. Because of the high operating temperature of the bulb wall, it is imperative that the lamp is not touched during installation. The dirt deposited impairs the light transmission and the bulb overheats at the point touched, leading to premature failure. The light output is between 15 and 25 lumens per watt, according to rating. (*Figure 7.25*).

The mercury–metal halide lamp

This is similar to the high-pressure mercury discharge lamp except that small quantities of metallic halides are included in the envelope. Examples of these are sodium halide and thallium halide. When the mercury discharge attains a sufficiently high temperature, the metals are liberated from their compounds in the same way as in the tungsten halide lamp. In this case, however, the metals are not deposited, but stay in a free form in the mercury discharge. When an electrical discharge takes place in any element the colour produced is unique, as we have already seen in the case of the sodium lamp. The metals included therefore determine the colour of the discharge. The light output from the mercury discharge together with the additional colours from the metals gives an output very close to that of sunlight and this lamp is used extensively to light stadia. The light output is 70 lumens per watt. As an example, a large stadium from which colour television transmissions are to take place would achieve comparable results to daylight using 240 lamps each of 1600 W give a level of surface illumination of 1400 lux. The lamp construction is similar to that of the high-pressure mercury type,

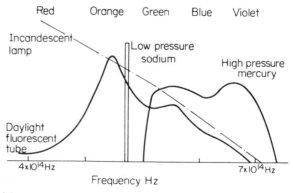

Figure 7.26

except that there is no auxiliary electrode; starting is by high-voltage impulse, as in the sodium lamp.

The high-pressure sodium lamp

If the pressure in the sodium lamp is increased and the temperature raised to about 1000°C, the discharge becomes much whiter and blends well with light from normal incandescent lamps. This development could not take place until a new material for the tube had been developed. The low-pressure lamp uses a glass tube while this lamp uses an aluminium oxide (alumina) tube which is capable of withstanding the very high temperature. It is used extensively for floodlighting where the warm, slightly pink colour is very attractive. The light output is about 100 lumens per watt.

TYPES OF LUMINAIRE

A selection of luminaires available is shown in *Figure 7.27*.

Types (a) to (d) are used indoors, types (e) to (i) outdoors. They can be diffusing or concentrating and are chosen for these functions, at the same time considering their appearance and the environment in which they are to be used. Outside luminaires must be weather-proof

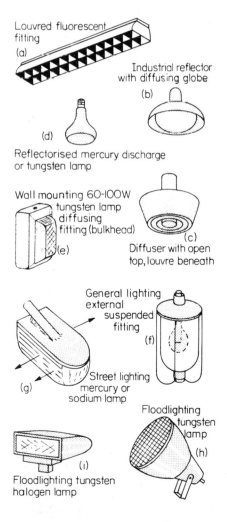

Figure 7.27

and as far as possible, vandal-proof. The transparent parts are made of prestressed glass or impact resistance plastic. To minimise maintenance, the surfaces should be self-cleaning by the action of wind and rain. Therefore any patterns or prisms are on the inner surface of the glass. Electrical connections to them are generally by means of conduit or mineral insulated cables.

DAYLIGHTING

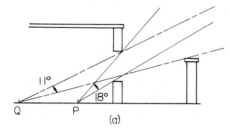

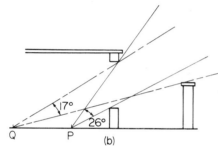

Figure 7.28

In most buildings use is made of daylight when available, supplemented by artificial light. The correct balance between the two is difficult especially when the rooms are deep when the variation in daylighting from a position near a window to that several metres from a window may be considerable.

The amount of daylight reaching a point inside a building through a window directly from the sky will depend on the area of the sky visible from that point, on the luminance of the sky and the angle at which it arrives on the surface to be lit. As already stated:

$$E = \frac{I}{d^2} \cos \theta° \text{ lux.}$$

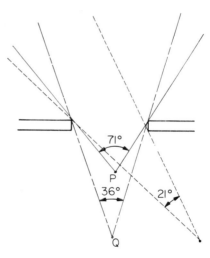

Figure 7.29

Figure 7.28 shows how the arc of sky visible is dependent on the distance inside the room. In *Figure 7.28(a)* at point P there is 18° of sky visible whilst at point Q there is only 11°. Note how the wall some distance from the building restricts the light input to point Q. If this wall were not there the arc would extend downwards to the first obstruction which might be the window sill or the horizon. In *Figure 7.28(b)* the effect of a taller window can be seen. More arc of sky is visible from both points. Note also in both cases the difference in the angle of incident light between points P and Q.

Figure 7.29 shows a plan view of the room. Note how the arc of sky visible is limited in the horizontal plane also. At points off the centre line the arc is further reduced.

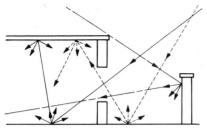

— Light entering the room directly from the sky

—·— Light entering the room by reflection from an external vertical surface

--- Light entering the room by reflection from an external horizontal surface

Figure 7.30

The luminance of the sky depends on the season of the year, the hour of the day and the distance of the room from the equator, i.e. on its latitude, since the height of the sun above the horizon depends on these factors. The average level of illumination on a horizontal surface in the open, which is a measure of the luminance of the sky, measured at Greenwich using north light only varies from about 450 lux at 9 a.m. to 1700 lux at midday in January compared with 4150 lux and 5700 lux at the same times in June. In addition, the luminance of the sky will depend on whether it is overcast or not. This may be due to clouds or pollution.

One further factor affecting the amount of light entering a room is the direction in which it is facing, north-facing rooms receiving no direct sunlight. Finally there is a component of light derived by reflection from external surfaces and this depends on the state and colour of those surfaces. This is illustrated in *Figure 7.30*.

Once light has entered a room by whatever means, the final distribution will depend on the state and colour of the decorations, diffused reflection taking place on most surfaces. *Figure 7.31* shows a typical illuminance curve for a room lit by north light.

Figure 7.31

PROBLEMS FOR SECTION 7

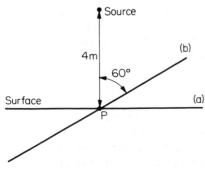

Figure 7.32

9 The light source in *Figure 7.32* has a uniform luminous intensity of 64 cd. The point P on the surface to be illuminated is 4 m from the source. Calculate the illuminance at point P when the surface is in position (a) and then in position (b).

10 Why are incandescent light bulbs often made from obscured (pearl) glass?

11 Why is the tungsten planar filament suitable for slide and cinema projectors?

12 Why does it take approximately 10 minutes for the high-pressure mercury vapour lamp to reach its full brilliance?

13 What would happen to a low-pressure sodium discharge lamp if its vacuum envelope were removed?

14 Why is it that the high pressure sodium lamp has been available only in the last few years?

15 What advantages has the tungsten halide lamp in the field of public lighting and automobile engineering?

16 What special features should a luminaire have if it is to be used outdoors?

17 List the ways in which daylight reaches a table top in a room with its only window in the vertical plane.

18 Upon what factors does the illuminance at a surface in a room depend when lit by daylight only?

19 A roadway is illuminated by means of lanterns which have luminous intensities in the vertical plane as stated in the table. The standards are 5 m high and 40 m apart.

Angle from the vertical Degrees	Luminous intensity Candela
0	300
10	310
20	330
30	360
40	400
50	450
60	530
65	590
70	685
75	810
80	660
85	0

Calculate the illuminance due to the nearest two luminaires at a point on a line on the ground joining two standards, 10 m from one of them.

20 A room 20 m × 40 m is equipped with 50 fluorescent tubes each giving a total output of 3400 lumens when new. The coefficient of utilisation for the room is 0.56 and a depreciation factor of 1.25 has been allowed. Calculate the level of surface illuminance (a) when all new tubes have just been installed and (b) just before these lamps are due for complete replacement.

21 An indoor floor area 30 m × 50 m is to be lighted to an illuminance of 100 lux. The building has an overall utilisation factor of 0.6. A depreciation factor of 1.1 is to be allowed.
Consider two schemes:

(a) using high-pressure sodium lamps each of which has a light output of 12 000 lumens when run in for a power input of 130 W.

(b) using 500 W incandescent lamps with an efficiency of 14 lumens per watt.

For each case:
 (i) calculate the required number of lamps and suggest a pattern of lamps to suit,
(ii) calculate the cost of energy per annum if the lamps are switched on for 3000 hours and the electricity tariff is £20/kW of maximum demand + 4 p/kWh of energy used.

8 Fuse protection

Aims: At the end of this chapter you should be able to:

Explain the action of circuit protection using fuses.
Describe the construction of common types of fuse.
Define fusing factor.
Explain how a fuse limits the energy input to a circuit.
Define arcing and pre-arcing time and sketch a typical current–time graph for a fuse.

FAULT LEVEL

When considering circuit protection, the fault level at the point to be protected must be taken into account. The fault level is the number of volt-amperes which would flow into a short circuit if one occurred at the particular point. On a 415 V three-phase street main this could be around 2000 kVA, which represents a current of nearly 3000 A. The fault level is limited by the impedance of all the apparatus between the point of the fault and the supply generator. When such a current flows a protective device such as a fuse should operate to cut off the supply. The value of current which could flow if nothing was done to limit it is called the 'prospective current'.

At a similar voltage within a factory where the distance from the supply transformer may be less and the transformer itself larger, the fault level could rise to 18 000 MVA, which represents a current of approximately 25 000 A. The nearer is the fault to the supply generators, the larger is the fault level. A fault on the 400 kV system can give fault levels in excess of 30 000 MVA.

Earthing

Both the IEE Regulations and Electricity Regulations require one point on a transformer-fed system to be earthed. Such earthing provides a return path for earth current, so facilitating clearance of faulty circuits.

Power transformer secondaries are star connected and are earthed in exactly the same way as the generators (see *Figure 8.1*). A fault usually exists on a piece of equipment because the live conductor or part of the winding of the equipment comes into contact with the metal casing of the equipment. If this casing is not connected to earth then the equipment becomes potentially dangerous. A person standing on the ground touching the casing of the equipment will receive an electric shock as the potential difference between that of the casing and earth causes a current to flow through the body. It only requires a few milliamperes along a route including the heart to cause death. Earthing the equipment allows current to flow through the earth connection back to the supply transformer neutral. If the impedance of the earth loop is small, sufficient current will flow to melt the fuse in the supply line.

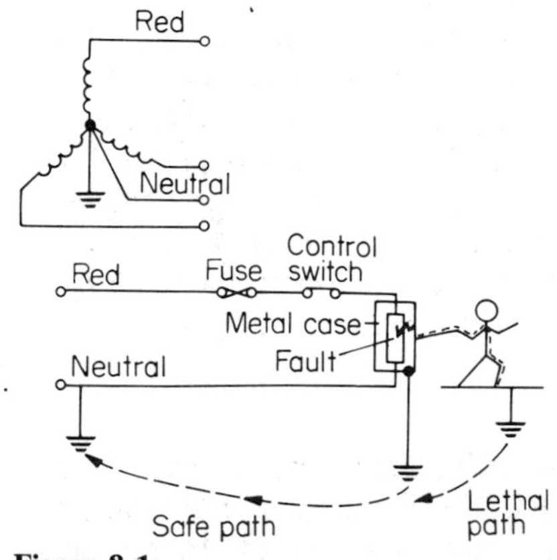

Figure 8.1

A high earth loop impedance is dangerous since the metal case will remain connected to the supply if insufficient current flows to cause fuse clearance.

In certain situations equipment is run without an earth wire. Such equipment is completely enclosed in a layer of insulating material or so disposed in an insulated box so that it is impossible for the box to become alive or for a person to push a finger in and so touch live metal. A television set in a plastics or wooden cabinet is an example of this type of construction.

FUSE PROTECTION

Two types of fuse are in general use:

1 the semi-enclosed, rewirable fuse
2 the high rupturing capacity fuse, or cartridge fuse (HRC).

Marked on each cartridge fuse and quoted for each size of fuse wire is a current rating. The current rating is that value of current which the fuse can carry indefinitely without melting or deteriorating.

The rated minimum fusing current is the least value of current which will actually cause the fuse element to melt.

$$\text{Fusing factor} = \frac{\text{Rated minimum fusing current}}{\text{Current rating}}$$

This factor may lie between approximately 1.2 and 2.0. With a fusing factor of 1.5, at fuse rated at 10 A will require a current of $1.5 \times 10 = 15$ A to operate.

OPERATION OF FUSES

When current flows in a fuse element, heat is produced. If that current is less than the current rating of the fuse then the body of the fuse is able to dissipate the heat to the surroundings, usually into the air. When the rate of producing heat is equal to the rate of dissipating it, then the temperature remains constant at a value lower than that necessary to melt the fuse element.

When a current slightly greater than the minimum fusing current flows, heat is produced in the element faster than it can be dissipated; after a period of time, the element melts and the circuit is interrupted. In the event of a short circuit or possibly an earth fault, the prospective current is extremely large and heat is generated in the element at such a high rate that there will be no time for all of it to be dissipated. The element melts extremely rapidly, circuit clearance times being very short. When operating in this manner the fuse is said to have 'inverse time characteristics'. This means that the larger the current, the shorter is the time taken to clear the circuit.

THE SEMI-ENCLOSED REWIREABLE FUSE

This comprises a ceramic fuse base with an asbestos or similar liner into which the fuse carrier is slotted. The fuse has male contacts of brass and pushes into spring-loaded female contacts of brass or berylium copper in the base. The fuse element is tinned-copper wire, the

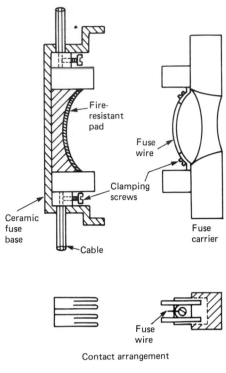

Fire-resistant pad

Fuse wire

Clamping screws

Ceramic fuse base

Cable

Fuse carrier

Fuse wire

Contact arrangement

Semi-enclosed rewireable fuse

Figure 8.2

diameters for particular currents being specified in the IEE Regulations. These fuses are the subject of BS 3036. (See *Figure 8.2*.)

The fuses are manufactured with ratings up to about 100 A and are suitable for use in alternating-current circuits operating at up to 250 V. In clearing, the fuse element melts and so interrupts the circuit current. The copper of the element partly vaporises and partly spreads itself over the surface of the carrier. It operates perfectly satisfactorily with overloads and fairly small fault currents. However, with very high prospective currents this melting takes place explosively and current may continue to flow in the metallic vapour present. The danger here is that the fault would have to be cleared by a much larger protective device farther back in the network, possibly at the main circuit-breaker. Should this happen, a large part of the network would be shut down and many consumers left without a supply of energy. Rewireable fuses are also open to abuse in that they can be rewired using almost any conductor, such as a hairpin or a nail. The IEE Regulations recommend that wherever possible a cartridge-type (HRC) fuse be used. At the fault level found in domestic situations, a semi-enclosed rewireable fuse is generally quite adequate.

THE HIGH RUPTURING CAPACITY (HRC) FUSE

The HRC fuse consists of a cylindrical ceramic body to the ends of which are attached brass or copper end-caps. The fusible element is generally of silver, since this is the best known electrical conductor.

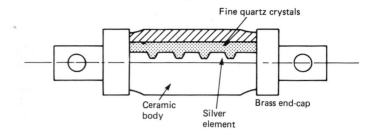

Fine quartz crystals

Ceramic body

Silver element

Brass end-cap

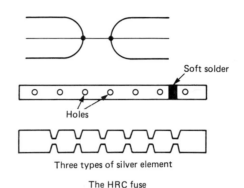

Soft solder

Holes

Three types of silver element

The HRC fuse

Figure 8.3

Three forms of the element are shown in *Figure 8.3*. On one of them a small bead of soft solder has been deposited. The body is tightly packed with very fine quartz grains which keep the element in place at all times, including whilst melting, and gives controlled heat conduction away from the element. With currents just large enough to cause melting, the element with the solder bead melts first at this point, whilst the other type melts at one or more of the narrow sections. The presence of the solder bead gives a lower fusing factor than is possible by most, if not all, other methods.

Where very large currents are involved, the heat produced, explosively melts the silver which chemically combines with the filling to form a compound, the resistance of which increases rapidly as it cools. BS 88 covers fuses for working up to 1000 V a.c. with ratings of several hundred amperes. They may be used to protect circuits with extremely high prospective currents, tests having been carried out on some types using values of over 100 kA. Cartridge fuses for domestic premises, generally only up to 100 A rating, are covered by BS 1361.

The small cartridge fuses found in domestic ring-main plugs have ratings between 3 A and 13 A and can interrupt up to 6000 A. They comprise a ceramic tube, quartz-filled with a single wire element, and employ the solder bead principle to give rapid clearance at low values of overcurrent.

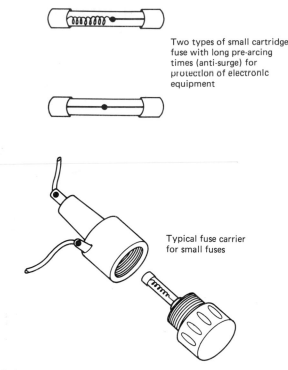

Two types of small cartridge fuse with long pre-arcing times (anti-surge) for protection of electronic equipment

Typical fuse carrier for small fuses

Figure 8.4

Small cartridge fuse links for telecommunication and light electrical apparatus are rated from about 50 mA to 5 A at 250 A. They are covered by BS 2950. Other miniature fuses appear in BS 4265.

Miniature fuses vary in speed from 'super quick-acting' to 'super time-lag'. The quick-acting types have a single-strand wire element of silver, copper or nickel–chromium, often with a small part of its length having reduced cross-sectional area. The time-lag varieties allow a surge to pass without rupturing and so are termed 'anti-surge' types. Two of these are shown in *Figure 8.4*. One of them has a single wire with a soft-solder bead situated about halfway along. With small prolonged overcurrents, the wire melts at the solder bead. With large overcurrents, the whole wire vaporises. In the other, with spring-assisted break, at low overcurrents the soldered joint melts, allowing the spring to pull back to give a good wide break. With severe overcurrents the straight section of the element (called the 'heater wire') vaporises and, again, the spring rapidly increases the gap. These types are not quartz-filled.

FUSE CHARACTERISTICS

When a much larger current than normal full load is allowed to flow in a circuit, the cable may suffer damage in two ways. Firstly, the excess current will generate a large amount of heat. Secondly, currents flowing in adjacent conductors set up mechanical forces ('Circuit damage due to overcurrents', Section 4.) These forces may be large enough to disrupt the cable.

For these reasons, the fuse must be capable of interrupting extremely large prospective currents before they ever reach their

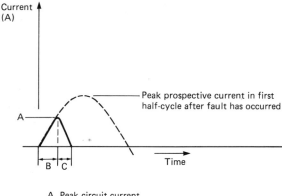

A Peak circuit current
B Pre-arcing time
C Arcing time

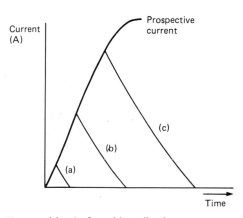

Clearance (a)   by fuse with small rating
          (b)   by fuse with medium rating
          (c)   by fuse with large rating

Figure 8.5

first maximum value. The fuse must begin to melt during the first quarter-cycle of current and so prevent the full prospective value being realised. *Figure 8.5* shows the prospective current in a circuit as a dotted line. This is the current which would flow in the circuit if the fuse were not present and nothing was done to limit the fault current. After a short time the energy dissipated in the fuse element by the large and rapidly increasing current is sufficient to start it melting. The time taken up to this instant is called the 'pre-arcing time'. Once the element begins to melt, the resistance of the fuse rapidly increases and the circuit current decreases. During this time there is an arc within the fuse and this period is called the 'arcing time'. The pre-arcing time plus the arcing time is the total clearance time.

Each manufacturer makes a large range of fuses, each with a different characteristic to suit a particular requirement. It is only

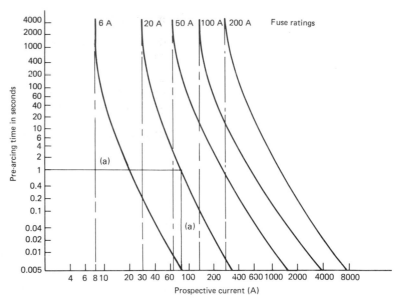

Figure 8.6

possible to deal with these in general terms. To determine the precise time for clearance of a particular fault, the actual fuse manufacturer's data must be consulted. *Figure 8.6* shows a general set of inverse time characteristics for a fuse. Looking at the curve for the 20 A fuse, for example, it may be seen that it will carry almost 30 A for ever without clearing. This tells us that its fusing factor is in the region of 30/20 = 1.5. As faults with increasing prospective currents are thrown onto the fuse, the pre-arcing times decrease. Following the lines marked (a) in the figure we see that, with a prospective fault current of 80 A, the fuse has a pre-arcing time of 1 second. This is after 50 complete cycles of the current. Proceeding down the curve to its intersection with the bottom axis, it is seen that with a prospective current slightly in excess of 300 A the pre-arcing time is 0.005 seconds, which is one quarter of a cycle. For prospective currents greater than 300 A the pre-arcing times will be even shorter, total clearance being achieved in less than one half-cycle.

We have been considering the fuse as a device which limits the size of a potentially very large current or limits the time during which a current only marginally greater than normal full load value is allowed to flow. Another way of viewing the fuse's action is to consider the amount of energy being allowed into the protected circuit. If we consider a 240 V supply feeding a load through a cable, the total resistance of the circuit (including that of the fuse) being 0.05 Ω, in the event of a perfect short-circuit at the end of this cable the current could theoretically rise to a value 240/0.05 = 4800 A. If this current flows for a complete cycle the energy input to the circuit can be calculated as follows:

Power input = I^2R = $4800^2 \times 0.05 = 1.152 \times 10^6$ W

Energy = power × time. Time for one cycle = 0.02 s (at 50 Hz).

Energy $= 1.152 \times 10^6 \times 0.02 = 23\,040$ joules.

Now suppose that the resistance of the fuse itself is 0.01 Ω and that of the circuit is 0.04 Ω. Consider the quantity I^2t.

$I^2t = 4800^2 \times 0.02 = 460\,800$ A^{2}s. Since this quantity does not contain the resistances of the two parts of the circuit, it applies both to the fuse and to the cable.

Energy used in clearing the fuse $= I^2Rt = (I^2t) \times R = 460\,800 \times 0.01$
$= 4608$ joules.
Notice that this is the actual energy used in melting the fuse element.

Energy supplied to the cable $= I^2Rt = (I^2t) \times R = 460\,800 \times 0.04$
$= 18\,432$ joules.

Total energy supplied $= 18\,432 + 4608 = 23\,040$ joules as before.

The fuse manufacturer could tell us the actual energy requirement to melt the fuse, but it is just as useful to quote the I^2t value since this is the same for both fuse and circuit. The value quoted assumes that there is no time for the element to pass heat to the surrounding air, i.e. the value quoted is for short-circuit fault conditions.

Fuse manufacturers quote values of I^2t input (i) to cause arcing to commence and (ii) to clear the fuse totally; provided that the I^2t requirement to clear the fuse is less than that necessary to damage the associated circuit, all is well.

This is particularly important in high-power electronics employing devices such as thyristors. These can carry extremely high currents but are physically very small. The amount of energy required to destroy them is quite small and a special range of very fast HRC fuses has been developed to protect such devices.

As an example, consider two 20 A fuses: (i) a high-speed type having I^2t value of 250 A^{2}s and (ii) a motor protection fuse with $I^2t = 3000$ A^{2}s. In each case the circuit being protected develops a short-circuit fault with resistance 0.05 Ω.

Fuse (i) will allow $I^2t \times 0.05 = 250 \times 0.05 = 12.5$ joules into the circuit, whereas fuse (ii) will allow $3000 \times 0.05 = 150$ joules to pass.

Which fuse is selected will depend upon knowing the energy level which will damage the circuit. If it is a cable, 150 joules is unlikely to harm it. If the fuse is connected directly to a solid semi-conductor device, it may well be that it will be necessary to limit the energy input to 12.5 joules to prevent damage.

**PROBLEMS FOR
SECTION 8**

1 A fuse has a rating of 15 A and a fusing factor of 1.3. What is the minimum value of current which will cause the fuse to operate?

2 Why is it necessary to have a low earth loop impedance when depending upon a fuse to give earth leakage protection?

3 A circuit protected with a 20 A HRC fuse will have a smaller conductor size than a circuit protected by a 20 A semi-enclosed fuse. Why is this? (See also Section 4, 'Effect of ambient temperature on rating'.)

4 Why is it necessary to interrupt fault currents before they rise to their full prospective value?

5 Sketch a curve showing 'pre-arcing time', 'arcing time' and 'total clearance time' for a fuse.

6 Sketch the construction of: (i) an HRC fuse for a power circuit, (ii) a cartridge fuse for a piece of low-rated electronic equipment and (iii) a semi-enclosed fuse for use on domestic premises.

7 What does the term 'anti-surge' mean with reference to small cartridge fuses?

8 Why are extremely fast-acting HRC fuses used in power semi-conductor circuits?

9 A fuse manufacturer quotes a value of I^2t of 0.8×10^6 A²s for the total clearance of an HRC fuse. A fault occurs on a circuit being protected by this fuse such that the circuit resistance falls to $0.02\ \Omega$. How much energy will enter the circuit whilst the fuse is clearing?

10 What is the necessary relationship between total clearance I^2t of a fuse and the permitted level of I^2t of the protected equipment, to give satisfactory protection?

Solutions to problems

SECTION 1

4 (i) 285.7 A, 16.33 kW; (ii) 28.57 A, 163 W.

18 19.05 kV.

19 216.5 V.

21 0.972 p.u.

22 AB, 23.8 A, BC 3.81 A, CD 26.2 A, Min. P.D. at C = 218.43 V.

23 50 A load power = 11.794 kW.

24 AB 89.6 A, BC 69.6 A, CD 39.6 A, DE 29.6 A, EF 20.4 A, FG 40.4 A, GA 65.4 A. Min. P.D. at E = 235.9 V.

25 (a)   50 A load; 231.75 V; 11.587 kW.
      125 A load; 230.12 V; 28.765 kW.
       80 A load; 235.22 V; 18.817 kW.

  (b)   Efficiency = 59.169/61.293 = 0.965 p.u.

SECTION 3

2 2.68 p/k Wh. Load factor 0.076

4 M.D. = 228.3 kW. 4.256 p/kWh.

6 £2222.22. £1422 after paying for correcting equipment.

7 (a)(i) 0.183, (ii) 0.363, (b) 0.0476, (c) 2.384 p/kWh, £29.80.

8 (a) 4.3 p/kWh, (b) 4.54 p/kWh.

9 £8000.

11   (i)   167.7 kVA, 0.894 p.f.
    (ii) 230.77 kVA, 175.4 kVA$_r$.
    (iii) 294.06 kW, 420.1 kVA.

12 (a) (i) 781.25 kVA,   (ii) 0.64
   (b) (i) 640.3 kVA,   (ii) 0.78.
   (c) (i) 538.5 kVA,   (ii) 0.928.
   (d) 500 kW = 500 kVA. (ii) 1.0 (unity p.f.)

13 £18 000–£12 048.80 = £5951.18 saving.

14 (a) 323.44 kW, 462.06 kVA. (b) to save £2000, kVA = 362.06. Required kVA$_r$ = 167.3.

15 (a) £25 756.30.
   (b) M.D. saving £1821.22. Cost £1024. Overall saving £778.8.
   (c) M.D. saving £473.48. Cost £737. Not sound.

16 (a) £15 000 + £52 560 = £67 560.
   (b) New M.D. = 450 kW.   New kWh = 1379 700
       New bill £64 188.   Saving £3372.

SECTION 4

4 (a) 1.5 mm^2, (b) 2.5 mm^2, (c) 6 mm^2, (d) 1.5 mm^2 (ambient temperature factor requires 14.08 A rating), (e) 2.5 mm^2.

9 (i) 1.08 V, (ii) 1.74 V, (iii) 2.75 V.

10 (a) 1.5 mm^2, (b) 1.5 mm^2, (c) 4 mm^2, (d) 16 mm^2.

11 10 kW, 1000 N, both per metre run.

13 X_c = 318 310Ω, I_R = 3.3 mA, I_C = 0.207 A, δ = 0.91°. Power loss in the dielectric = 217.8 W.

20 0.875 A.

21 (a) 71.4 A/mm², (b) 18.87 A/mm².
26 (i) Cold-rolled grain-oriented steel, more expensive per tonne but higher permeability means less weight and overall cost. Silicon content keeps eddy current losses low.
   (ii) Ordinary soft iron, cheap, not so efficient but limited duty time.
   (iii) Dust core or ferrite.
   (iv) Air.
27 In presence of strong alternating field.
28 (a) Mumetal or Stalloy. (b) Aluminium or copper.
29 150 W.
30 300 W.

SECTION 5

4 80.
6 (a) 112.5 V, (b) 2.778 A, (c) 187.5 W.
15 (a) 0.67, (b) 0.31.
19 (i) 5.208 A, (ii) 48 V, (iii) 0.8 power factor.

SECTION 6

3 (a) 180 V, (b) 360 V, Power = 14 400 W in both cases.
8 (a) 504 V, (b) 29.14 rev/s.
9 (a) 1008 V, (b) 14.57 rev/s.
14 (a) 129 V, (b) 360 V.
15 26 rev/s ($E = 510$ V, k$\Phi = 19.6$).
16 (a) 10.83 rev/s, (b) 0.036 Wb.
19 (a) Add 9.7 Ω, (b) add 4.7 Ω.
21 Output = 19 150 W, efficiency = 0.837 p.u.
22 Input = 25 832 W, efficiency = 0.852.
24 Speed control down to a crawl with minimum losses.

SECTION 7

4 2948 cd, 37 047 lm.

7 $\frac{250}{500} \cos 63.4 + 2 = 2.224$ lx.

9 (a) 4 lx, (b) 3.46 lx.

19 From the nearer lamp, $E = \frac{590}{5^2 + 10^2} \cos 65.3 = 1.97$ lx.

From the further lamp, $E = \frac{600}{5^2 + 30^2} \cos 80.6 = 0.106$ lx.

Total illuminance = 2.076 lx.
20 New light flux = 95 200 lm $E = 119$ lx. At the end, $E = 95.2$ lx.
21 (a) (i) 22.9 lamps (use 24). 4 rows of 6.
   (b) (i) Each lamp gives 7000 lm, 39.28 lamps (use 40). 5 rows of 8.
   (a) (ii) M.D. £62.40.   Energy £374.40.   Total £436.80.
   (b) (ii) M.D. £400.   Energy £2400.   Total £2800.

SECTION 8

1 19.5 A.
9 16 000 joules.

Index